RS403 .C667 2004

Contemporary drug
  synthesis /
      c2004.

                    2005 09 26

D0931331

Contemporary Drug Synthesis

Contemporary Drug Synthesis

Jie Jack Li

Douglas S. Johnson

Drago R. Sliskovic

Bruce D. Roth

PROPERTY OF
SENECA COLLEGE
LIBRARIES
@ YORK CAMPUS

WILEY-
INTERSCIENCE

A JOHN WILEY & SONS, INC., PUBLICATION

Copyright © 2004 by John Wiley & Sons, Inc. All rights reserved

Published by John Wiley & Sons, Inc , Hoboken, New Jersey
Published simultaneously in Canada

No part of this publication may be reproduced, stored in a retrieval system, or transmitted in any form or by any means, electronic, mechanical, photocopying, recording, scanning, or otherwise, except as permitted under Section 107 or 108 of the 1976 United States Copyright Act, without either the prior written permission of the Publisher, or authorization through payment of the appropriate per-copy fee to the Copyright Clearance Center, Inc , 222 Rosewood Drive, Danvers, MA 01923, (978) 750-8400, fax (978) 646-8600, or on the web at www copyright com. Requests to the Publisher for permission should be addressed to the Permissions Department, John Wiley & Sons, Inc , 111 River Street, Hoboken, NJ 07030, (201) 748-6011, fax (201) 748-6008

Limit of Liability/Disclaimer of Warranty While the publisher and author have used their best efforts in preparing this book, they make no representations or warranties with respect to the accuracy or completeness of the contents of this book and specifically disclaim any implied warranties of merchantability or fitness for a particular purpose. No warranty may be created or extended by sales representatives or written sales materials. The advice and strategies contained herein may not be suitable for your situation You should consult with a professional where appropriate. Neither the publisher nor author shall be liable for any loss of profit or any other commercial damages, including but not limited to special, incidental, consequential, or other damages.

For general information on our other products and services please contact our Customer Care Department within the U S at 877-762-2974, outside the U S. at 317-572-3993 or fax 317-572-4002

Wiley also publishes its books in a variety of electronic formats Some content that appears in print, however, may not be available in electronic format.

Library of Congress Cataloging-in-Publication Data is available.

ISBN 0-471-21480-9

Printed in the United States of America

10 9 8 7 6 5 4 3 2 1

Preface

Armed with a vast amount of knowledge acquired at school, along with a degree in chemistry, you are starting your career in the pharmaceutical industry. You may be working in medicinal chemistry, process chemistry, or radiochemistry. This new endeavor may seem daunting, especially when you consider that you may need to be proficient in areas in which you have not been prepared. This manuscript will illustrate how chemistry, biology, pharmacokinetics, and a host of other disciplines all come together to produce successful new medicines. In order to achieve that goal, we have compiled a collection of fourteen representative categories of drugs that we have carefully chosen from among the best-selling drugs. We have provided an introduction to each drug including a historical perspective, and background to the biology, pharmacology, pharmacokinetics, and drug metabolism followed by a detailed account of the synthesis. The targeted audience goes beyond individuals new to the pharmaceutical industry; many veterans of the industry may well find something new in this text. There are a few points we felt worth reiterating. For example:

a). The advent of new synthetic methodology enables chemists to synthesize drugs in a more convergent and more efficient fashion, a theme seen over and over again in this monograph. Consequently, literature awareness is essential for the chemist in the pharmaceutical industry.

b). Knowing the history of successful drugs and understanding their attributes is very important and the lessons learned can be applied to current programs in drug discovery. A tremendous amount of knowledge has accumulated over the last few decades. Attributes of a successful drug include appropriate potency, selectivity, bioavailability, and physiochemical properties.

c). Serendipity in the drug industry also plays an important role. On the other hand, opportunity favors prepared minds. Many examples can be found in this manuscript. For example, Viagra®, currently used for erectile dysfunction (ED), was initially developed as a PDE5 inhibitor for hypertension. Likewise, Rogaine®, currently used topically for hair growth, was first synthesized as a potassium channel opener, also for hypertension. Propecia®, currently used orally for hair growth, was originally prepared as a testosterone-5α-reductase inhibitor for the treatment of benign prostatic hyperplasia (BPH). Therefore, one should be cognizant during the clinical trials; even failure for the initially intended therapeutic indication does not necessarily equate the end of a drug as long as it has been proven to be safe.

d). Another point that medicinal chemists often overlook (sometimes justifiably) is that the synthesis of the drug should be eventually amenable to cost-effective scale-up to make it economically viable. This is where process chemists play an important role. In the subsequent chapters, we have incorporated process synthesis routes when the information was available.

We are grateful to Professor David L. Van Vranken at University of California at Irvine, who read the manuscript and offered many valuable comments and suggestions. We are also indebted to Dr. Alan J. Kraker, Dr. Lorna H. Mitchell, Dr. Derek A. Pflum, Dr. Stephen Cho, Larry Bratton, and William C. Patt for proofreading portions of the manuscript.

Jie Jack Li

Douglas S. Johnson

March 2004

Table of Contents

VIII

Brand Names and Their Corresponding USANs

Abilify®	aripiprazole
Accutane®	isotretinoin
Advair®	fluticasone propionate and salmeterol xinafoate
Allegra®	fexofenadine
Amerge®	naratriptan hydrochloride
Axert®	almotriptan malate
Celebrex®	celecoxib
Cialis®	tadalafil
Cipro®	ciprofloxacin
Clarinex®	desloratadine
Claritin®	loratadine
Crestor®	rosuvastatin
Finasteride®	finasteride
Flonase®	fluticasone propionate
Flovent®	fluticasone propionate
Frova®	frovatriptan succinate
Geodon®	ziprasidone
Gleevec®	imatinib mesylate
Imitrex®	sumatriptan succinate
Lescol®	fluvastatin
Levitra®	vardenafil hydrochloride
Lipitor®	atorvastatin calcium
Maxalt®	rizatriptan benzoate
Mevacor®	lovastatin
Nexium®	esomeprazole
Paxil®	paroxetine hydrochloride
Plavix®	clopidogrel
Prilosec®	omeprazole
Relpax®	eletriptan hydrobromide
Risperdal®	risperidone
Pravacol®	pravastatin
Prozac®	fluoxetine hydrochloride
Rogaine®	minoxidil

Serevent®	salmeterol xinafoate
Seroquel®	quetiapine fumarate
Singulair®	montelukast sodium
Tazorac®	tazarotene
Ticlid®	ticlopidine
Viagra®	sildenafil citrate
Vioxx®	rofecoxib
Xenical®	orlistat
Zocor®	simvastatin
Zoloft®	sertraline hydrochloride
Zomig®	zolmitriptan
Zyprexa®	olanzapine
Zyrtec®	cetirizine dihydrochloride
Zyvox®	linezolid

Acronyms and Abbreviations

Ac ....acetyl
ADP ....adenosine diphosphate
ALIQUAT .... tricaprylmethyl ammonium chloride
cAMP .... adenosine cyclic 3',5'-phosphate
ATP .... adenosine triphosphate
AUC .... area under curve
BER...borohydride exchange resin
BINAP.. 2,2'-bis(diphenylphosphino)-1,1'-binaphthyl
BMI .... body mass index
Boc ....*tert*-butyloxycarbonyl
Bn .... benzyl
BPH ....benign prostatic hyperplasia
BSTFA .... bis(trimethylsilyl)-trifluoroacetamide
t-Bu..*tert*-butyl
CL..total clearance
CL$_R$ ....renal clearance
CML .... chronic myeloid leukemia
CNS .... central nervous system
COX-2 ....cyclooxygenase II
m-CPBA .... *m*-chloroperoxybenzoic acid
CSF... cerebal synovial fluid
5-CT 5-carbamoyltryptamine
CYP .... cytochrome
DABCO.. 1,4-diazabicyclo[2.2.2]octane
DALYS ....Disability Adjusted Life Years
DBU .... 1,8-diazabicyclo[5.4.0]undec-7-ene
o-DCB .... *o*-dichlorobenzene
DCC... 1,3-dicyclohexylcarbodiimide
DDQ.. 2,3-dichloro-5,6-dicyano-1,4-benzoquinone
DMF... dimethylformamide
DMSO...dimethylsulfoxide
DHT 5α-dihydrotestosterone
DNA ....deoxy-nucleic acid
ECG...ecocardiograms
ED .... erectile dysfunction
EGF .... epidermal growth factor
EPS .... extrapyramidal side-effects
FDA...Food and Drug Administration
Fen–Phen...fenfluramine and phentermine
GI...gastrointestinal
GISA .... glycopeptide-intermediate *S. aureus*

CGMP .... cyclic guanosine monophosphate
GPCRs.. G-protein-coupled receptors
HMG-CoA.. hydroxymethylglutaryl coenzyme A
HMGR...HMG-CoA reductase
HMPA .... hexamethylphosphoric triamide
HPL.. human pancreatic lipase
HPLC..high-performance liquid chromatography
5-HT ....5-hydroxytryptamine (scrotonin)
KCO ....potassium channel opener
LAH.. lithium aluminum hydride
LDA.. lithium diisopropylamide
LHMDS ....lithium hexamethyldisilazane
LTs ....leukotrienes
MAO ....monoamine oxidase
MDD ....major depressive disorder
MICs ....minimal inhibition concentrations
MMPP .... magnesium monoperoxyphthalate hexahydrate
MOA ....mechanism of action
MRSA .... methicillin-resistant *Staphylococcus aureus*
NBS .... *N*-bromosuccinimide
NCS ....*N*-chlorosuccinimide
NSAIDs ....non-steroidal anti-inflammatory drugs
OA .... osteoarthritis
PCC .... pyridinium chlorochromate
PDE5 .... phosphodiesterase-5
PDGFR.. platelet-derived growth factor receptor, a kinase
PG.. prostaglandin
PK...pharmacokinetics
PKC...protein kinase C
PLE...pig liver esterase
PPH .... primary pulmonary hypertension
PPI...proton pump inhibitor
RA .... rheumatoid arthritis
Ra-Ni ....Raney-Nickel
RCM..ring-closing metathesis
RNA .... ribonucleic acid
RT...room temperature
SDAs .... serotonin-dopamine antagonists
S_NAr .... nucleophilic substitution on an aromatic ring
S_N1.. unimolecular nucleophilic substitution
S_N2.. bimolecular nucleophilic substitution
SNRI.. serotonin and noradrenaline reuptake inhibition
SPOS .... solid phase organic synthesis
SSRI's .... selective serotonin reuptake inhibitors
T .... testosterone
Tbf ....tetrabenzo[*a,c,g,i*]fluorene

TBS ....*tert*-butyldimethylsilyl
TCAs ....tricyclic antidepressants
TEA...triethylamine
TES...triethylsilyl
Tf.. trifluoromethanesulfonyl (triflyl)
TFA ....trifluoroacetic acid
TFAA .... trifluoroacetic anhydride
THF .... tetrahydrofuran
TKI .... tyrosine kinase inhibitor
TMEDA...*N,N,N',N'*-tetramethylethylenediamine
TMG.. tetramethylguanidine
Tol ....toluene or tolyl
Ts...tosylate
USAN...United States Adopted Names
UV ....ultraviolet
VRE.. vancomycin-resistant enterococci
V_{ss} .... steady-state volume of distribution

Chapter 1. Antithrombotics:

Ticlopidine (Ticlid®) and Clopidogrel (Plavix®)

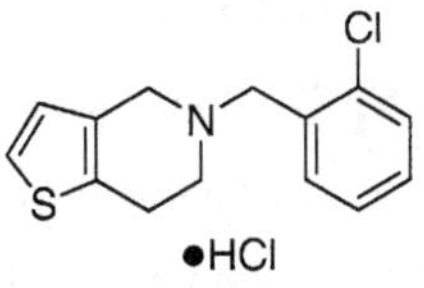

USAN· Ticlopidine hydrochloride
Trade Name: Ticlid ®
Castaigne S. A.
Launched: 1979
M.W. 263.6

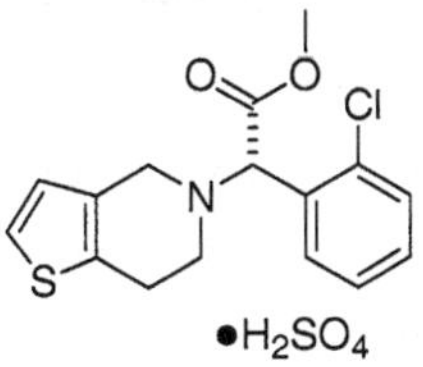

1

USAN: Clopidogrel sulfate
Trade Name: Plavix®
Sanofi/Bristol–Myers Sqibb
Launched: 1993
M.W. 321.8

2

§1.1   Introduction[1–8]

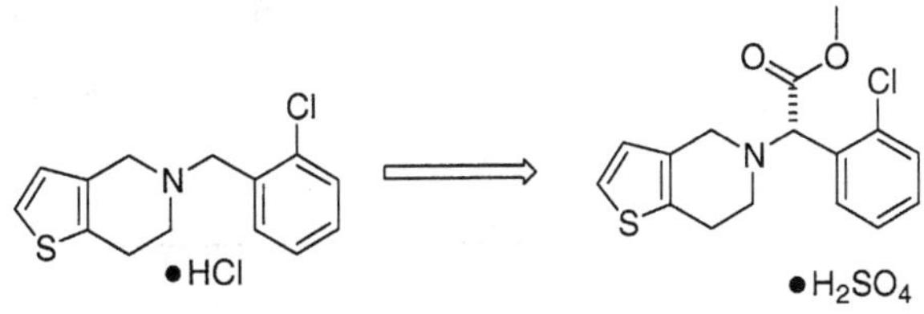

1, Ticlopidine   2, Clopidogrel

Fig. 1. Evolution of ticlopidine (**2**) to clopidogrel (**1**).

The process of thrombosis involves the aggregation of platelets, resulting in a pro-coagulation state in the blood that may form a blood clot in the vascolature. Both ticlopidine (**1**) and clopidogrel (**2**) inhibit platelet aggregation induced by adenosine diphosphate (ADP), a platelet activator that is released from red blood cells, activated platelets, and damaged endothelial cells. ADP also induces platelet adhesion. Ticlopidine (**1**) is an older drug, launched in 1979, whereas clopidogrel (**2**), launched in

1993, has achieved with great commercial success. Like its predecessor ticlopidine (**1**), clopidogrel (**2**) is an ADP-dependent platelet aggregation inhibitor. The mechanisms of action for both ticlopidine (**1**) and clopidogrel (**2**) are the same — through the antagonism of the P2Y12 purinergic receptor and prevention of binding of ADP to the P2Y12 receptor. However, both ticlopidine (**1**) and clopidogrel (**2**) are not active *in vitro*, but are activated *in vivo* by cytochrome P450-mediated hepatic metabolism. Remarkably, the identity of the active metabolite **5** of clopidogrel (**2**) was unknown until 1999,[9] when it was isolated after exposure of clopidogrel (**2**) or 2-oxo-clopidogrel (**4**) to human hepatic microsomes. It is noteworthy that ticlopidine (**1**) and clopidogrel (**2**) do not share a common active metabolite.

Due to the aforementioned metabolic activation, an induction period is required for the metabolism-based threshold buildup when patients start to take clopidogrel (**2**); therefore, it may take some time for the effect to manifest.

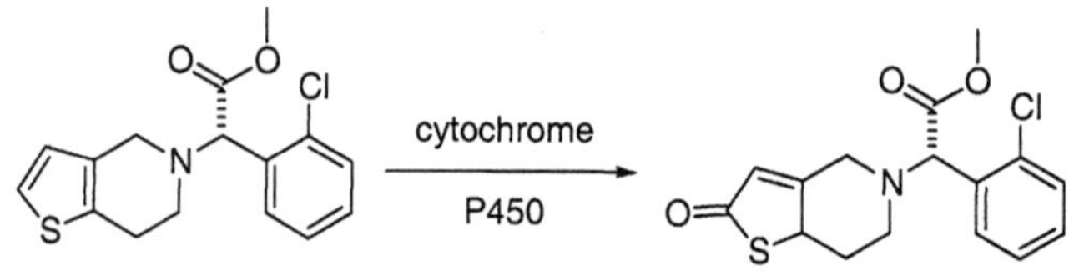

2, clopidogrel
inactive *in vitro*

4, 2-oxo-clopidogrel
inactive *in vitro*

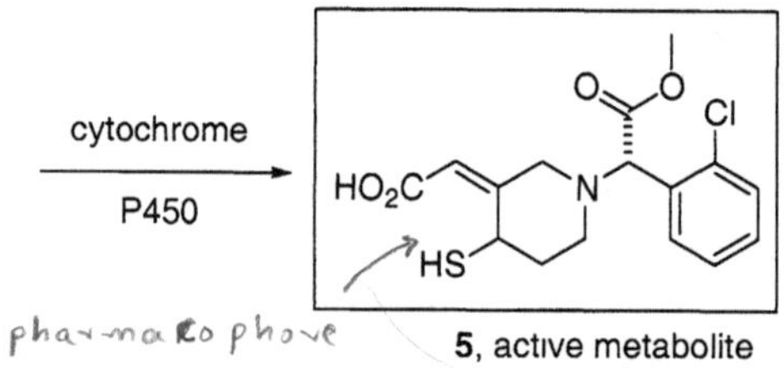

5, active metabolite
active *in vitro*

Fig. 2. *In vivo* metabolism of clopidogrel (**1**).

§1.2 Syntheses of ticlopidine (1)[10–12]

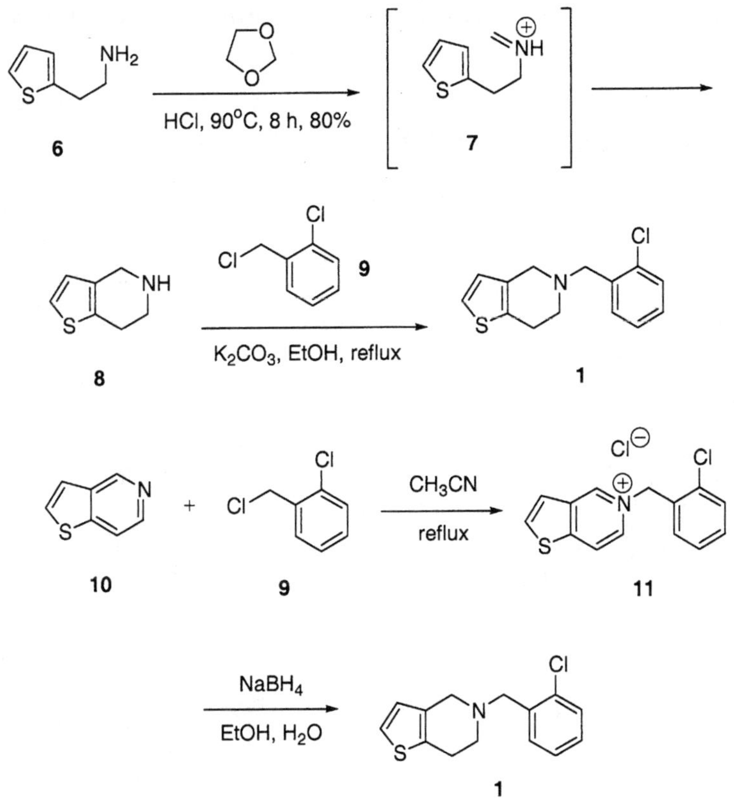

Scheme 1. Synthesis of ticlopidine (1).

In one route, ticlopidine (1) was assembled *via* S_N2 displacement of 2-chlorobenzyl chloride (9) with 4,5,6,7-tetrahydro-thieno[3,2-c]pyridine (8).[10,11] The nucleophile 8 was synthesized by heating 2-thiophen-2-yl-ethylamine (6) with 1,3-dioxolane in the presence of concentrated hydrochloric acid.[12] 1,3-Dioxolone gave better yields than with formaldehyde, paraformaldehyde and 1,3,5-trioxane. The interesting transformation 6 → 8 first involved the formation of the corresponding Mannich base 7, which then underwent a Pictet–Spengler type reaction to afford the ring-closure product 8. It was of interest to note that a possible intramolecular aminomethylation did not take place.

Another route toward ticlopidine (**1**) involved an S$_{N}$2 displacement of 2-chlorobenzyl chloride (**9**) by thieno[3,2-c]pyridine (**10**) to produce thieno[3,2-c]pyridinium chloride **11**. Subsequent reduction of the pyridinium salt (**11**) using NaBH$_4$ then delivered ticlopidine (**1**).[12]

§1.3   Syntheses of Clopidogrel (2)[13–19]

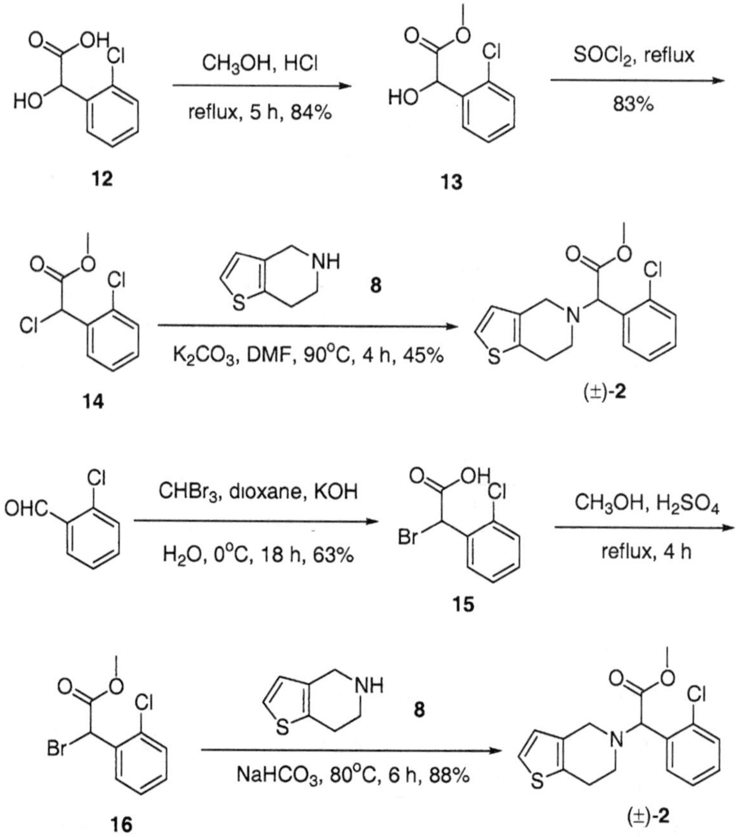

Scheme 2. The Sanofi syntheses of racemic (±)-clopidogrel (**2**).

The original Sanofi synthesis of (±)-clopidogrel (**2**) began with the formation of the methyl ester **13**. Thus methyl mandelate **13** was prepared by refluxing chlorinated mandelic acid **12** with methanol in the presence of concentrated HCl.[13] Chlorination of

13 using thionyl chloride gave methyl α-chloro-(2-chlorophenyl)acetate (**14**). S$_N$2 displacement of **14** with thieno[3,2-c]pyridine (**8**) then delivered (±)-clopidogrel (**2**).

Alternatively, α-bromo-(2-chlorophenyl)acetic acid (**15**) was prepared by treatment of 2-chlorobenzaldehyde with tribromomethane in dioxane with an aqueous solution of potassium hydroxide.[14] Formation of methyl ester **16** was followed by an S$_N$2 displacement by thieno[3,2-c]pyridine (**8**) to afford) an 88% yield of (±)-clopidogrel (**2** in two steps.

One of the disadvantages of the synthetic route delineated in Scheme 1 is the use of 2-thiophen-2-yl-ethylamine (**6**), whose preparation was not only difficult, but also expensive. Recently, RPG Life Science revealed their one-pot synthesis of ticlopidine and (±)-clopidogrel that did not require the use of 2-thiophen-2-yl-ethylamine (**6**).[15] (2-Chlorobenzyl)-(2-thiophen-2-yl-ethyl)-amine hydrochloride (**17**) was prepared according to US 4127580 and US 5204469. Refluxing **17** with paraformaldehyde in *tert*-butanol with the aid of 37% concentrated HCl gave ticlopidine (**1**). Likewise, refluxing **18** with paraformaldehyde in *iso*-propanol with the aid of 37% concentrated HCl gave (±)-clopidogrel (**2**).

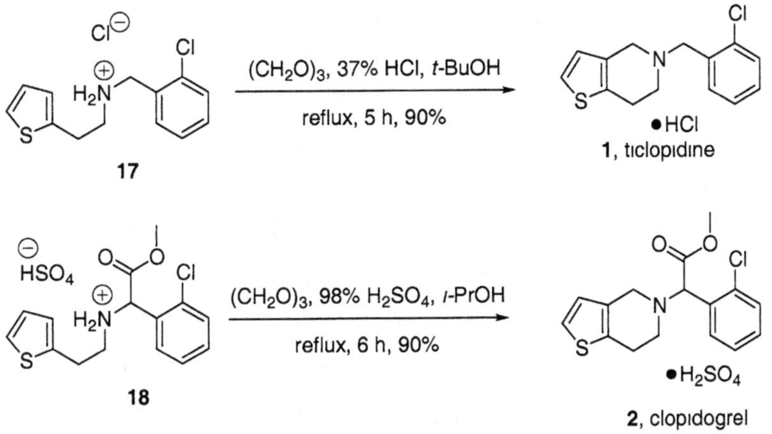

Scheme 3. The RPG synthesis of ticlopidine (**1**) and (±)-clopidogrel (**2**).

Interestingly, only the enantiomerically pure (+)-clopidogrel (**2**) exhibits platelet aggregation inhibiting activity, while the (−)-(**2**) is inactive.[16] Moreover, the inactive (−)-

(2) isomer is 40 times less well tolerated between the two enantiomers. Therefore, it is evidently advantageous to administer (+)-clopidogrel (2). With regard to the enantiomerically pure (+)-clopidogrel (2), it was originally obtained from resolution of the racemic clopidogrel (2) or through intermediates that were derived *via* resolution. For instance,[17] racemic clopidogrel (2) was treated with levorotary camphor-10-sulfonic acid in acetone to afford a salt 19, which was recrystallized from acetone to generate (+)-clopidogrel (2).

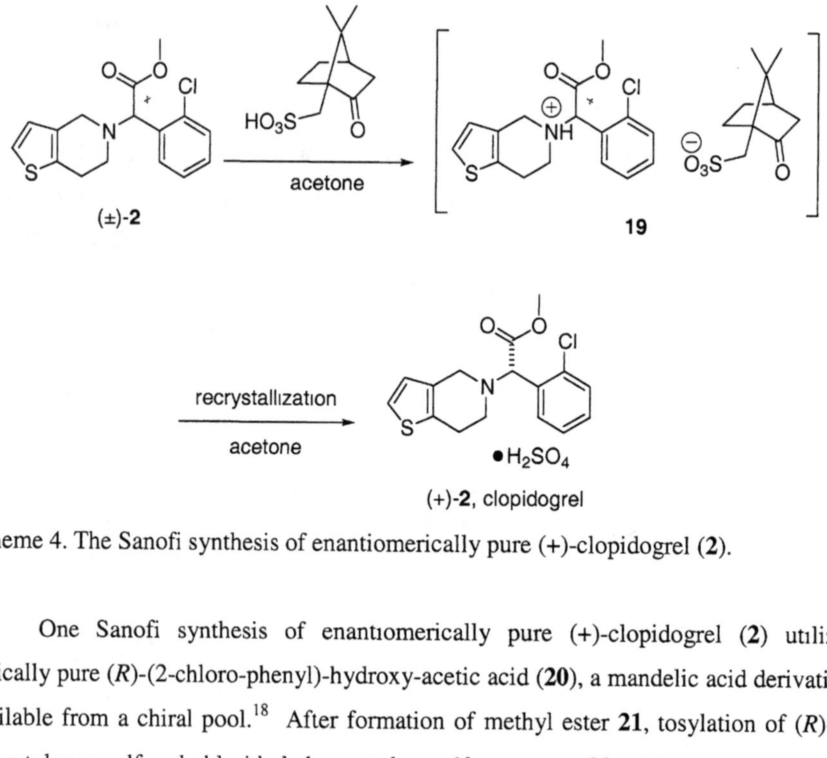

Scheme 4. The Sanofi synthesis of enantiomerically pure (+)-clopidogrel (2).

One Sanofi synthesis of enantiomerically pure (+)-clopidogrel (2) utilized optically pure (R)-(2-chloro-phenyl)-hydroxy-acetic acid (20), a mandelic acid derivative, available from a chiral pool.[18] After formation of methyl ester 21, tosylation of (R)-21 using toluene sulfonyl chloride led to α-tolenesulfonate ester 22. Subsequently, the S_N2 displacement of 22 with thieno[3,2-c]pyridine (8) then constructed (+)-clopidogrel (2). Another Sanofi synthesis of enantiomerically pure (+)-clopidogrel (2) took advantage of resolution of racemic α-amino acid 23 to access (S)-23.[19] The methyl ester 24 was prepared by treatment of (S)-23 with thionyl chloride and methanol. Subsequent S_N2 displacement of (2-thienyl)-ethyl *para*-toluene-sulfonate (25) assembled amine 26.

Finally, ring-closure was achieved by heating **26** with paraformaldehyde in formic acid at reflux to give (+)-clopidogrel (**2**).

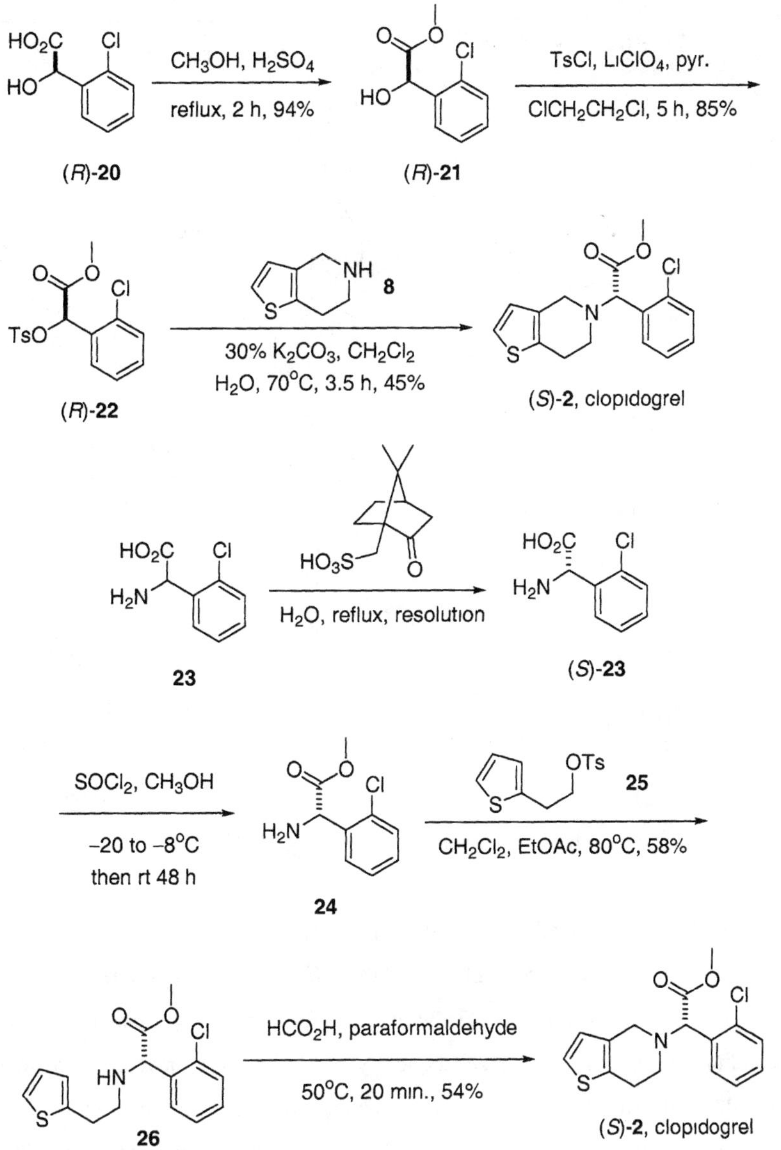

Scheme 5. Asymmetric synthesis of (+)-clopidogrel (**2**).

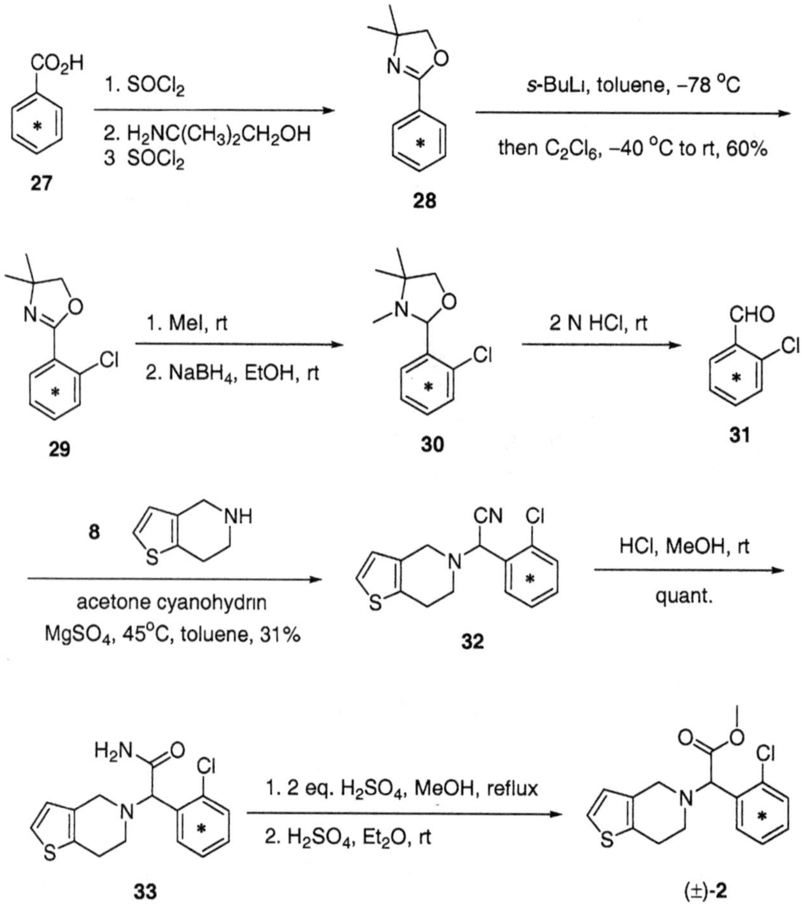

Scheme 6. The Sanofi synthesis of labeled (±)-clopidogrel (1).

A synthesis of labeled (±)-clopidogrel (2) has been described.[20] The synthesis commenced with the commercially available [*benzene*-U-^{13}C]-benzoic acid (27). Treating the acid chloride of 27 with 2-amino-2-methyl-propan-1-ol gave the corresponding amide, which was subsequently treated with thionyl chloride to afford oxazalıne 28. Applying Myers' oxazoline-directed *ortho*-lithiation methodology, 28 was exposed to *s*-butyllithıum and then quenched with hexachloroethane to give chloride 29 ın 60% yield after separation of unreacted starting material 28. Methylation of oxazoline

29 was followed by reduction of the resulting iminium intermediate to give trimethyl-oxazolidine **30**. Acidic hydrolysis of the cyclic hemiaminal functionality on **30** then produced [*benzene*-U-^{13}C]-2-chlorobenzaldehyde (**31**). A Strecker reaction on aldehyde **31** with 4,5,6,7-tetrahydro-thieno[3,2-*c*]pyridine (**8**) in the presence of acetone cyanohydrin and MgSO$_4$ in toluene assembled nitrile **32**. While saponification of **32** resulted in predominantly a retro-Strecker reaction, hydrochloric acid-promoted hydrolysis gave primary amide **33** in quantitative yield. Finally, methanolysis and salt formation proceeded uneventfully to deliver [*benzene*-U-^{13}C]-(±)-clopidogrel (**2**). The overall yield for this sequence from [*benzene*-U-^{13}C]-benzoic acid (**27**) to [*benzene*-U-^{13}C]-(±)-clopidogrel (**2**) was 7%.

§1.4   References

1.   Ticlopidine, (a) Saltiel, E.; Ward, A. *Drugs* **1987**, *100*, 1667–1672. (b) Castañer, J. *Drugs Fut.* **1976**, *1*, 190–193.

2.   MOA of Ticlopidine and Clopidogrel, Savi, P.; Labouret, C.; Delesque, N.; Guette, F.; Lupker, H.; Herbert, J. M. *Biochem. Biophysic. Res. Commun.* **2001**, *283*, 379–383.

3.   Ticlopidine and Clopidogrel, Quinn, M. J.; Fitzgerald, D. J. *Circulation* **1999**, *34*, 222–262.

4.   Clopidogrel, Jarvis, B.; Simpson, K. *Drugs* **2000**, *60*, 347–377.

5.   Clopidogrel, Herbert, J. M.; Fréhel, D.; Bernat, A.; Badorc, A.; Savi, P.; Delebasseee, D.; Kieffer, G.; Defreyn, G.; Maffrand, J. P. *Drugs Fut.* **1996**, *21*, 1017–1021.

6.   Clopidogrel, Escolar, G.; Heras, M. *Drugs of Today* **2000**, *36*, 187–199.

7.   Iqbal, O.; Aziz, S.; Hoppensteadt, D. A.; Ahmad, S.; Walenga, J. M.; Bakhos, M.; Fareed, J. *Emerging Drugs* **2001**, *6*, 111–135.

8.   Bennett, J. S. *Ann. Rev. Med.* **2001**, *52*, 161–184.

9.   Savi, P.; Pereillo, J. M.; Fedeli, O.; *et al. Thromb. Haemost.* **1999**, *82*, Suppl. 230, 723.

10.   Maffrand, J. P.; Eloy, F. *Eur. J. Med. Chem.* **1974**, *9*, 483–486.

11.   Maffrand, J. P.; Eloy, F. *J. Heterocyclic. Chem.* **1976**, *13*, 1347–1349.

12.   Sumita, K.; Koumori, M.; Phno, S. *Chem. Pharm. Bull.* **1994**, 42, 1676–1678.

13.   Eliel, E. L.; Fisk, M. T.; Prosser, T. *Org. Synth.* **1963**, *Coll. Vol. IV*, 169–172.

14.   Aubert, D.; Ferrand, C.; Maffrand, J.-P. US 4529596 (**1985**).

15.   Tarur, V. P.; Srivastava, R. P.; Srivastava, A. R.; Somani, S. K. WO 0218357 (**2002**).

16.   Bouisset, M.; Radison, J. US 5036156 (**1991**).

17.   Badorc, A.; Fréhel, D. US 48747265 (**1989**).

18.   Bousquet, A.; Musolino, A. WO 9918110 (**1999**).

19.   Descamps, M.; Radisson, J. US 5204469 (**1993**).

20.   Burgos, A.; Herbert, J. M.; Simpson, I. *J. Labelled. Compd. Radiopharm.* **2000**, *43*, 891–898.

Chapter 2. Anti-inflammatory Cyclooxygenase-2 Selective Inhibitors: Celecoxib (Celebrex®) and Rofecoxib (Vioxx®)

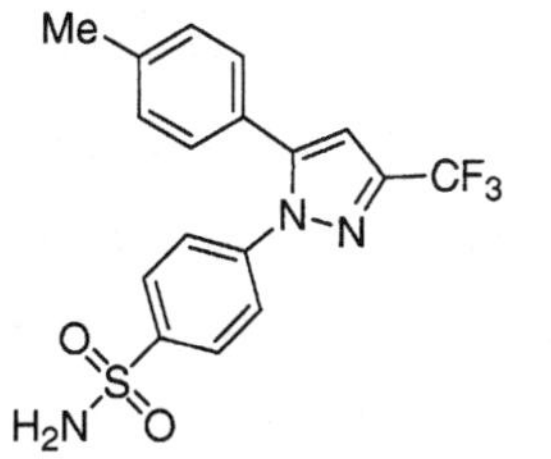
Me
CF3
N
N
O
S
H2N
O

1

USAN: Celecoxib
Trade Name: Celebrex®
Pfizer
Launched: 1998
M.W. 381.37

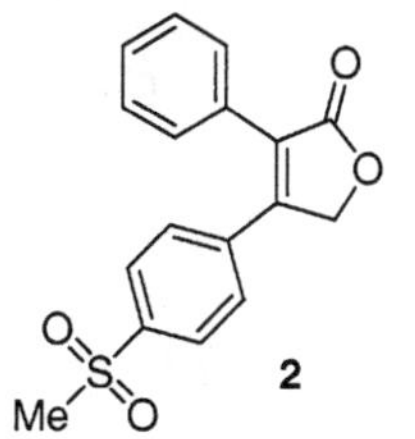
O
O
O
S
Me
O

2

USAN Rofecoxib
Trade Name: Vioxx®
Merck
Launched. 1999
M.W. 314.36

---

§2.1 Background

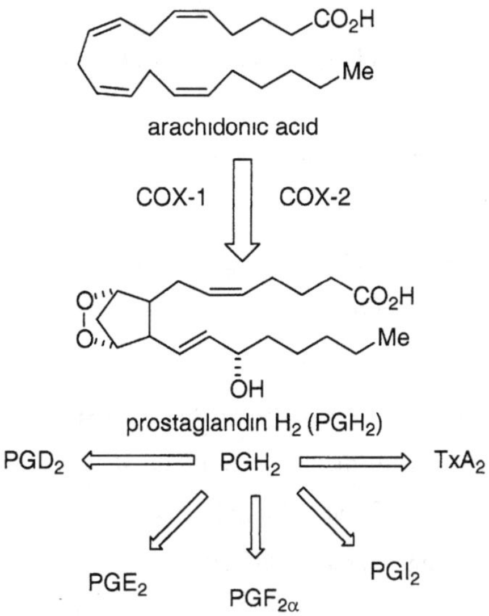
CO2H
Me
arachidonic acid
COX-1
COX-2
CO2H
Me
OH
prostaglandin H2 (PGH2)
PGD2
PGH2
TxA2
PGE2
PGF2α
PGI2

Fig. 1. The arachidonic acid cascade.

The arachidonic acid cascade (Fig. 1) plays an important role in the inflammation process. Inhibition of prostaglandin production, especially that of PGG_2, PGH_2 and PGE_2, has been a fertile approach for the discovery of anti-inflammatory drugs. A plethora of conventional anti-inflammatory, analgesic and anti-pyretic agents, known as non-steroidal anti-inflammatory drugs (NSAIDs), are widely used for managing inflammatory disorders such as osteoarthritis (OA) and rheumatoid arthritis (RA). The most prevalent categories of NSAIDs include salicylates as represented by aspirin (**3**), arylacetic acids as exemplified by indomethacin (**4**), and aryl propionic acid such as ibuprofen (**5**) and naproxen (**6**). All of the conventional NSAIDs are inhibitors of the COX enzymes. This enzyme catalyzes the transformation of arachidonic acid to prostaglandin H_2.

3, aspirin   **4**, indomethacin   **5**, ibuprofen   **6**, naproxen

In 1991, a novel isoform of cyclooxygenase in the arachidonic acid/prostaglandin pathway was discovered.[1-4] This inducible enzyme associated with the inflammatory process was named cyclooxygenase II (COX-2) or prostaglandin G/H synthase II. COX-2 is localized mainly in inflammatory cells and tissues, and becomes up-regulated during the acute inflammatory response; whereas COX-1 is mainly responsible for the production of homeostatic prostaglandins which are associated with gastrointestinal ulceration, perforation and hemorrhage. Therefore, it was envisioned that a COX-2 selective inhibitor would be beneficial in inhibiting prostaglandin production for reducing the adverse gastrointestinal and hematologic side effects. Celecoxib (**1**) and rofecoxib (**2**), both marketed in 1999, are both COX-2 selective inhibitors that are effective anti-inflammatory agents that exhibit significantly reduced gastrointestinal side effects. Their selectivities and some pharmacokinetics parameters are highlighted in Table 1 and Table

2, respectively. The data are not a direct comparison because they were taken from two separate publications.[5–6]

	COX-2 whole blood	COX-1 whole blood	Ratio	COX-2 whole cell	COX-1 whole cell	Ratio
Celecoxib (1)[6]				0.04	15	350
Rofecoxib (2)[5]	0.6	10	17	0.01	4.7	470
Indomethacin (4)[5]	0.5	0.2	0.4	0.03	0.02	0.7

Table 1. Cyclooxygenase inhibitory activities of celecoxib (1), rofecoxib (2) and indomethacin (4), IC_{50} (μM).

	C_{max} ng/mL	T_{max} hr	$T_{1/2}$ hr	AUC ng*hr/mL	F%	V_{ss}/F L	CL/F L/hr	V_{dss} L
Celecoxib (1)[6]	705	2.8	11.2			429	27.7	
Rofecoxib (2)[5]	207 ±111	2–3		3286 ±843	93			86–91

Table 2. Some pharmacokinetics parameters of celecoxib (1) and rofecoxib (2).

§2.2   Synthesis of Celecoxib

With regard to the synthesis of celecoxib (1), several routes were described in the 1995 patent by G. D. Searle[6-7] As shown in Scheme 1, dione 7 was prepared by the Claisen condensation of 4-methylacetophenone with ethyl trifluoroacetate in the presence of NaOMe in methanol under reflux. Subsequent diarylpyrazole formation from the condensation of dione 7 and 4-sulfonamidophenylhydrazine hydrochloride then delivered celecoxib (**1**). The corresponding chlorophenylpyrazolyl analog **9** (SC-236) was potent (IC_{50} = 0.01 μM for COX-2), selective (IC_{50} = 17.8 μM for COX-1), and efficacious. However, the plasma half-life was unacceptably long. The plasma half-life for celecoxib (**1**) was 10–12 hours, which was unacceptably long. Replacement of the chlorine atom with the methyl group accelerated the metabolism to the benzoic acid via the *in vivo* oxidation process.

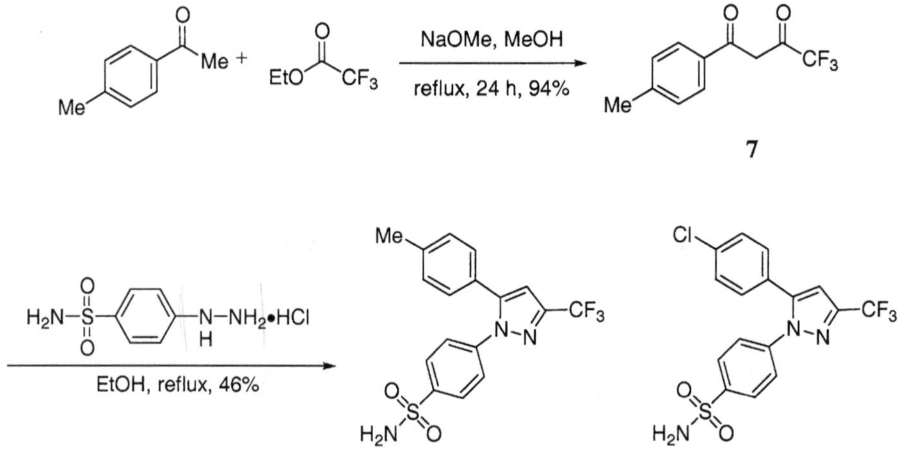

Scheme 1. Synthesis of celecoxib (**1**).

§2.3   Syntheses of Rofecoxib

In the original patent published by Merck in 1995,[8] rofecoxib (**2**) was synthesized in three steps from the known 4-(methylthio)acetophenone (**10**), prepared from the Friedel–Crafts acylation of thioanisole. As depicted in Scheme 2, oxidation of sulfide **10** using an excess of magnesium monoperoxyphthalate hexahydrate (MMPP, an inexpensive, safe and commercially available surrogate for *m*-CPBA) gave rise to sulfone **11**, which was subsequently brominated with bromine and $AlCl_3$ to afford 2-bromo-1-(4-(methylsulfonyl)phenyl)ethanone (**12**).[9] After recrystallization from 1:1 EtOAc/hexane, the pure phenylacyl bromide **12** was then cyclo-condensed with phenylacetic acid under the influence of 1,8-diazabicyclo[5.4.0]undec-7-ene (DBU) to deliver rofecoxib (**2**) in

58% yield. It is assumed that the ketoester from the initial esterification underwent an intramolecular Claisen condensation to furnish furanone **2**.

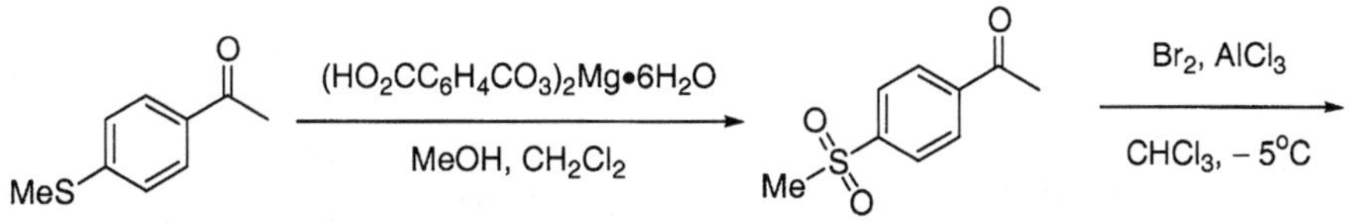

Scheme 2. The original synthesis using a cyclo-condensation strategy.

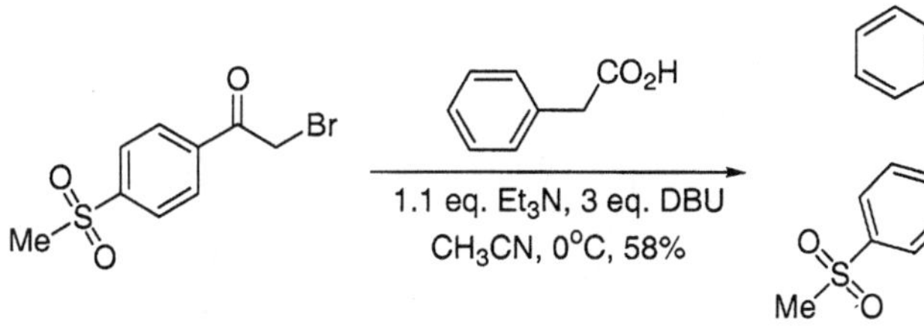

Subsequently, several variants of the synthesis were patented in 1996 and 1997, respectively.[10,11] In one approach, a Suzuki coupling strategy was employed for the installation of the second aryl group.[10] Under the action of Et₃N, phenylacetic acid and bromoacetate underwent an S_N2 reaction to afford diester **13** in excellent yield (Scheme 3). Phenyl tetronic acid **14** was obtained by treating diester **13** with a strong base such as KOt-Bu. Triflate **15**, generated from tetronic acid **14**, could be directly coupled with 4-methylthio-phenylboronic acid (**17**) to give furanone **18** [Pd(Ph₃P)₄, toluene, Na₂CO₃, EtOH, 60 °C, 70% yield]. Alternatively, triflate **15** was transformed to bromide **16**, which was subjected to the Suzuki coupling with boronic acid **17** in 90% yield. Boronic acid **17**, in turn, was prepared from 4-methylthio-phenylbromide via a halogen-metal exchange and quenching with B(Oi-Pr)₃. Finally, oxone oxidation of diarylfuranone **18** using the phase-transfer catalyst Bu₄NBr furnished rofecoxib (**2**).

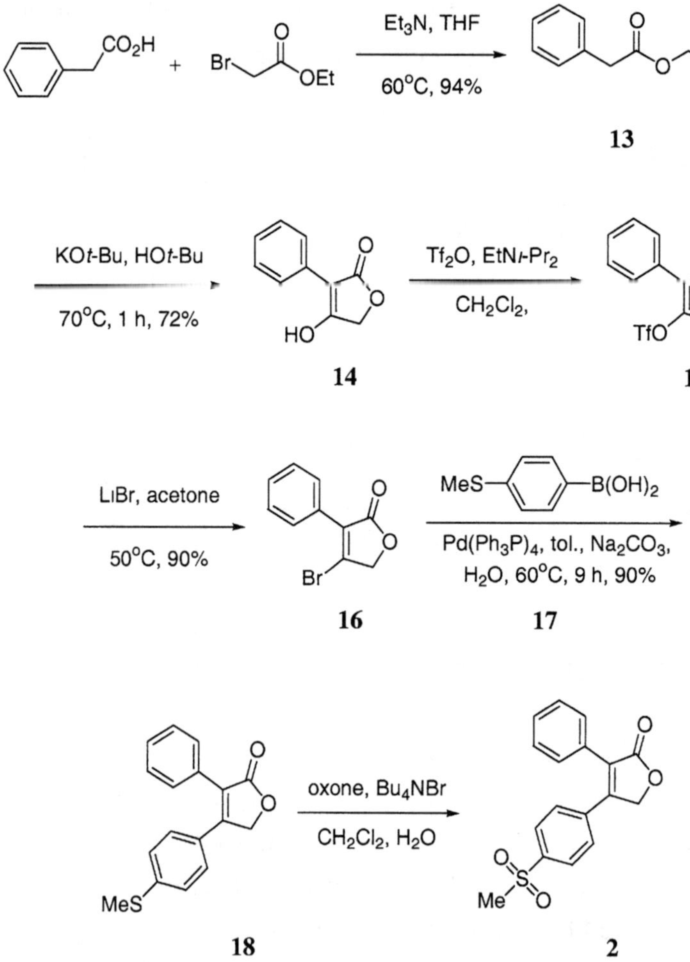

Scheme 3. The Suzuki coupling approach.

In a laboratory-scale synthesis of rofecoxib (**2**), a ruthenium-catalyzed lactonization of diarylacetylene **19** gave rise to **2** regioselectively (Scheme 4).[11] Diarylacetylene **19**, on the other hand, was prepared by utilizing a Castro–Stephens reaction of *p*-iodophenyl methyl sulfone and copper(I) phenylacetylene in pyridine under reflux.[12] In another case, Fallis and coworkers synthesized butenolide in **2** using a magnesium-mediated carbometallation reaction.[13] Therefore, treatment of phenylpropargyl alcohol with 4-methylthiophenylmagnesium chloride followed by oxidation with *m*-CPBA afforded **2** in quantitative yield.

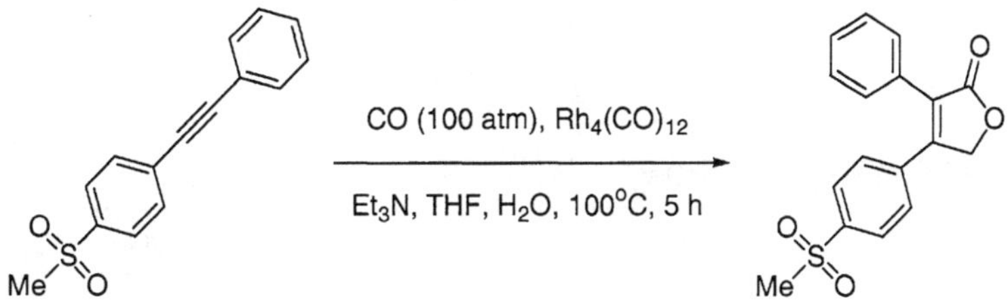

Scheme 4. The rhuthenium-catalyzed and magnesium-mediated furanone formation.

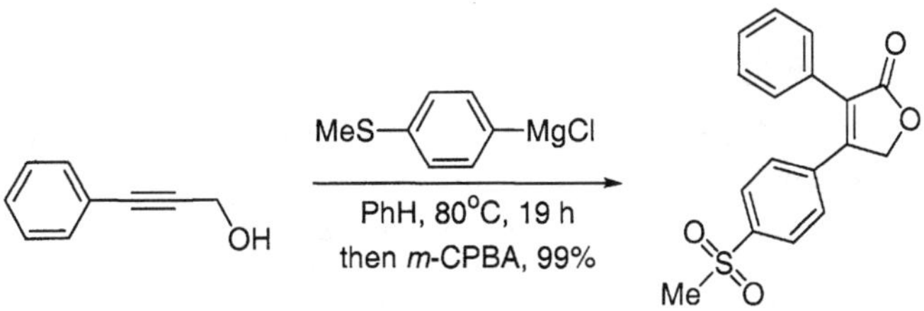

Interestingly, the synthesis of rofecoxib (**2**) on a process-scale[14] is reminiscent of the original route.[8]   Thus Friedel–Crafts acylation of thioanisole was performed employing the AlCl$_3$/CH$_3$COCl complex in *o*-dichlorobenzene (*o*-DCB) to furnish 4-(methylthio)acetophenone (**10**).   Without further purification, tan *o*-dichlorobenzene solution of **10** from the workup was subjected to oxidation using H$_2$O$_2$ and a catalytic amount of tungstate in the presence of ALIQUAT (tricaprylmethyl ammonium chloride, a phase-transfer catalyst) to produce keto-sulfone **11**.   The direct bromination of **11** using Br$_2$/HOAc initiated with HBr then provided bromoketone **12**.   Addition of **12** to a solution of sodium phenylactate, prepared *in situ* by reaction of phenylacetic acid with NaOH at 40°C, effected rapid coupling to give ketoester **20**.   Ketoester **20** was subsequently treated with diisopropylamine at 45°C in the same reaction vessel to deliver the desired rofecoxib (**2**).

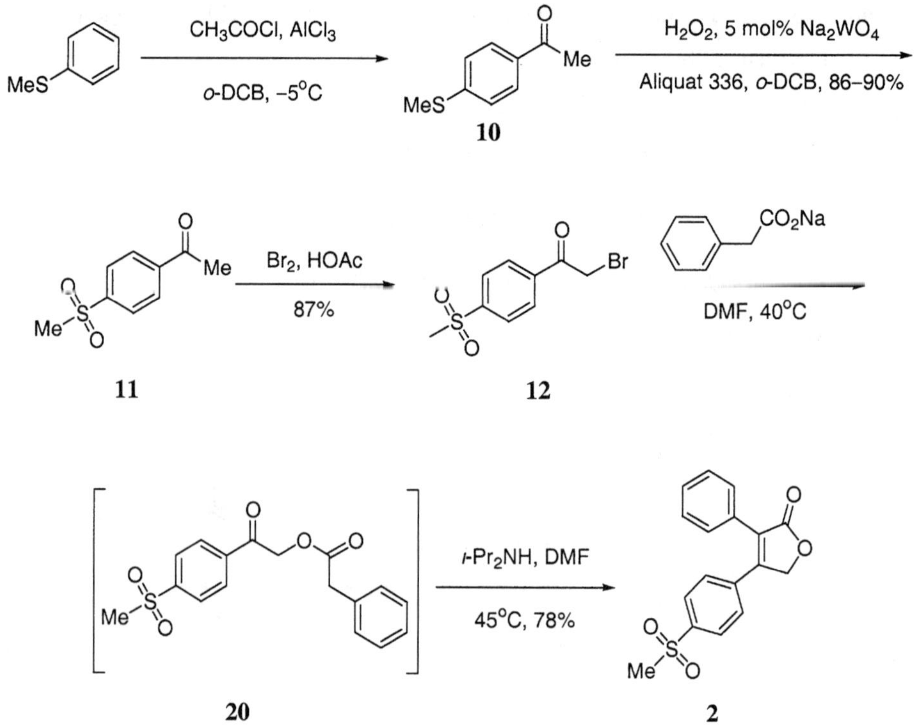

Scheme 5. Process-scale synthesis of rofecoxib (**2**).

*   *   *

The two COX-2 selective inhibitors, celecoxib (**1**) and rofecoxib (**2**), marketed in 1999 and 2000, respectively, have quickly become blockbuster drugs (with annual sales of more than one billion dollars) for the treatment of osteoarthritis (OA) and rheumatoid arthritis (RA). Many backups and competition drugs are sure to follow. Excitingly, some COX-2 selective inhibitors have been shown to inhibit cancer growth as well.[15]

§2.4   References

1.   *Clinical Significance and Potential of Selective COX-2 Inhibitors*, Vane, J.; Botting, R. eds; Williams Harvey Press, **1998**.

2.   Carter, J. S. *Expert Opinion on Therapeutic Patents* **2000**, *10*, 1011–1020.

3.   Marnett, L. J.; Kalgutkar, A. S. *Current Opinion in Chem. Biol.* **1998**, *2*, 482–490.

4.   Beuck, M. *Angew. Chem. Int. Ed.* **1999**, *38*, 631–633.

5.   Prasit, P.; Wang, Z.; Brideau, C.; Chan, C.-C.; Charleson, S.; Cromlish, W.; Ethier, D.; Evans, J. F.; Ford-Hutchinson, A. W.; Gauthier, J. Y.; Gordon, R.; Guay, J.; Gresser, M.; Kargman, S.; Kennedy, B.; Leblanc, Y.; Léger, S.; Mancini, J.; O'Neill, G. P.; Ouellet, M.; Percival, M. D.; Perner, H.; Riendeau, D.; Rodger, I.; Tagari, P.; Thérien, M.; Vickers, P.; Wong, E.; Xu, L.-J.; Young, R. N.; Zamboni, R. *Bioorg. Med. Chem. Lett.* **1999**, *9*, 1773–1778.

6.   Penning, T. D.; Talley, J. J.; Bertenshaw, S. R.; Carter, J. S.; Collins, P. W.; Docter, S.; Graneto, M. J.; Lee, L. F.; Malecha, J. W.; Miyahiro, J. M.; Rogers, R. S.; Rogier, D. J.; Yu, S. S.; Anderson, G. D.; Koboldt, C. M.; Perkins, W. E.; Seibert, K.; Veenhuizen, A. W.; Zhang, Y. Y.; Isakson, P. C. *J. Med. Chem.* **1997**, *40*, 1347–1365.

7.   Talley, J. J.; Penning, T. D.; Collins, P. W.; Rogier, D. J., Jr.; Malecha, J. W.; Miyashiro, J. M.; Bertenshaw, S. R.; Khanna, I. K.; Granets, M. J.; Rogers, R. S.; Carter, J. S.; Docter, S. H.; Yu, S. S. WO 95/15316.

8.   Ducharme, Y.; Gauthier, J. Y.; P.; Leblanc, Y.; Wang, Z.; Léger, S.; Thérien, M. WO 95/18799.

9.   Cutler, R. A.; Stenger, R. J.; Suter, C. M. *J. Am. Chem. Soc.* **1952**, *74*, 5475–5481.

10.   Desmond, R.; Dolling, U.; Marcune, B.; Tillyer, R.; Tschaen, D. WO 96/08482.

11.   Hancock, B.; Winters, C.; Gertz, B.; Ehrich, E. WO 97/44028.

12.   Eisch, J. J.; Hordis, C. K. *J. Am. Chem. Soc.* **1971**, *93*, 2974–2981.

13.   Forgione, P.; Wilson, P. D.; Fallis, A. G. *Tetrahedron Lett.* **2000**, *41*, 17–20.

14.   Desmond, R.; Dolling, U. H.; Frey, L. F.; Tillyer, R. D.; Tschaen, D. M. WO 98/00416.

15.   Marx, J. *Science* **2001**, *291*, 581–582.

16.   Review on celecoxib, Graul, A.; Martel, A. M.; Castañer, J. *Drugs Fut.* **1997**, *22*, 711–714.

17.   Review on rofecoxib, Sorbera, L. A.; Leeson, P. A.; Castañer, J. *Drugs Fut.* **1998**, *23*, 1287–1296.

Chapter 3. H⁺/K⁺-ATPase Inhibitors: Esomeprazole (Nexium®)

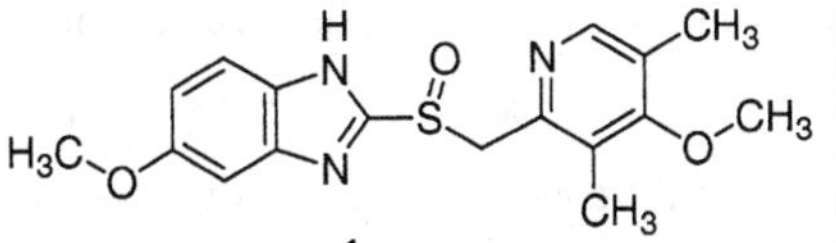

USAN: Omeprazole
Trade Name. Prilosec®
AstraZeneca
Launched:1985
M.W. 345 21

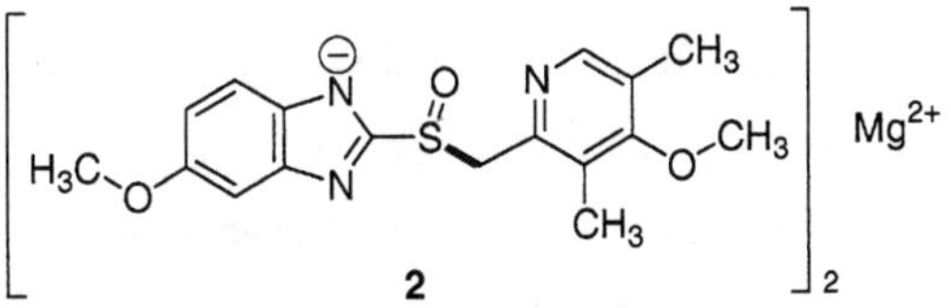

USAN. Esomeprazole
Trade Name Nexium®
AstraZeneca
Launched: 2001
M.W. 713.13

§3.1 Introduction

Hydrochloric acid (HCl), as the gastric acid, is a major constituent of the secretions of the gastrointestinal tract. It is essential to the food digestion process. However, excessive secretion of HCl will damage the mucosa and subsequently the submucosa, producing ulceration of the gastrointestinal tract, especially the stomach and duodenum. Thus maintaining the equilibrium between HCl and the acid-neutralizing mucus is important for the gastric system. Two types of drugs are utilized for neutralizing the excess acid produced in the GI tract: the first type are antacid salts, such as $CaCO_3$, $NaHCO_3$, and $AlPO_4$ whose mode of action is a simple acid-base neutralization; while the second type are inhibitors of acid production. The latter is evidently superior to the former due to lesser side effects. The inhibitor approach has produced many block-buster drugs that have revolutionized the pharmaceutical industry. Examples include H_2-blockers such as cimetidine (**3**) and ranitidine (**4**).

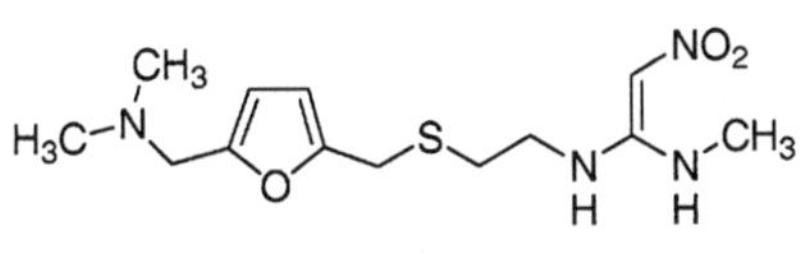

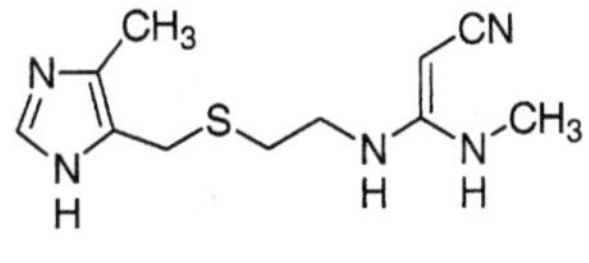

Fig. 1. Cimetidine (**3**) and ranitidine (**4**).

Omeprazole (**1**), introduced in 1988, was the first H^+/K^+-ATPase inhibitor, also known as a proton pump inhibitor (PPI), which was marketed as a treatment for gastric ulcers. It functions by preventing acid production in the mucosa. Omeprazole was the best-selling drug for several years until its patent expiration in 2001. The mechanism of action of omeprazole (**1**), the "omeprazole cycle", is shown in Scheme 1.[1,2] Consequently, protonation of **1** takes place slowly when it encounters the acidic medium in the gastric chamber. The protonated intermediate **5** then undergoes an intramolecular nucleophilic cyclization in a 5-*exo-dig* fashion where the pyridine moiety serves as the nucleophile to furnish benzoimidazoline **6**. A reversible ring opening of **6** then delivers sulfenic acid **7**, which serves as an electrophile in a 6-*exo-trig* ring-closure to afford the reactive intermediate **8** after dehydration. The pyridinium sulfydryl **8** is a very good electrophile, and is readily attacked by the cysteine residue of the enzyme H^+/K^+-ATPase. Therefore, omeprazole (**1**) behaves more like a pro-drug because pyridinium sulfydryl **8** is the actual inhibitory species.

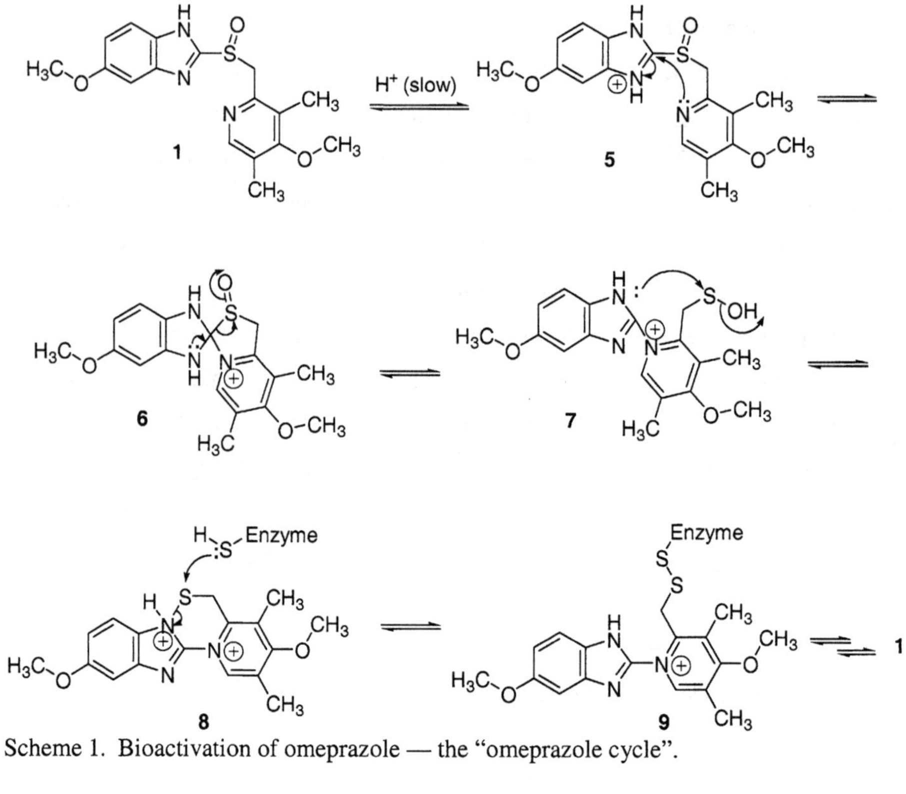

Scheme 1. Bioactivation of omeprazole — the "omeprazole cycle".

Esomeprazole (**2**) is the (*S*)-enantiomer of racemic omeprazole (**1**). The former has better pharmacokinetics and pharmacodynamics than the latter and therefore possesses higher efficacy in controlling acid secretion and has a more ideal therapeutic profile.

§3.2 Synthesis of Esomeprazole

The synthesis of esomeprazole (**2**) is illustrated below.

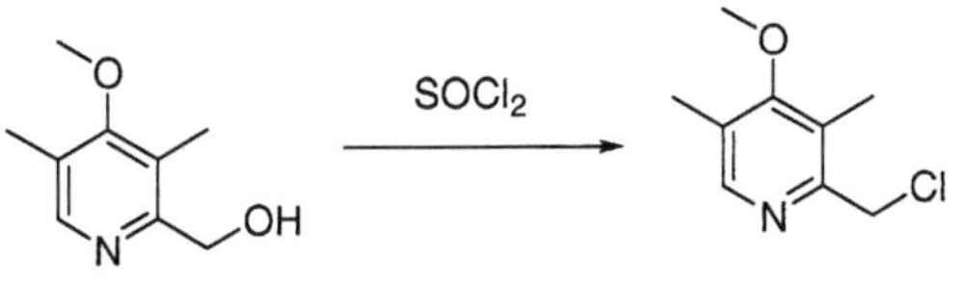

Pyridylmethyl chloride **11** was obtained by direct chlorination of alcohol **10** using thionyl chloride[3-5] The hydroxymethyl-pyridine **10**, in turn, was synthesized from 3,5-dimethylpyridine in 6 steps utilizing a Boekelheide reaction[6] that transformed the 2-methylpyridine-*N*-oxide to the corresponding hydroxymethyl-pyridine **10**.

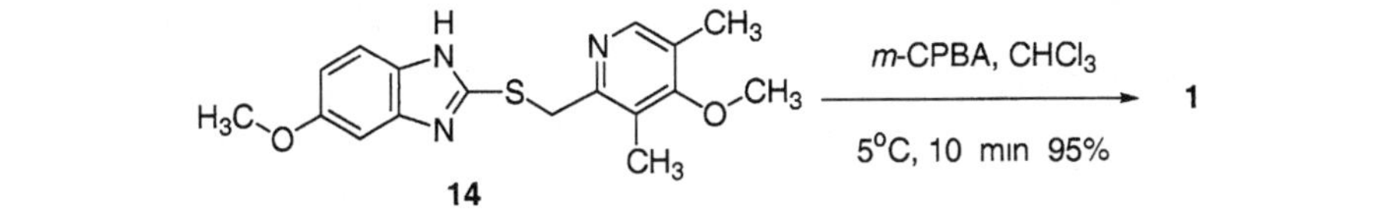

Methoxyphenylene diamine **12** was treated with potassium ethylxanthate to afford benzimidazole-thiol **13**. The coupling of thiol **13** and chloromethyl-pyridine **11** was then facilitated by treatment with NaOH in refluxing EtOH/H$_2$O. Subsequently, the oxidation of resulting sulfide **14** was easily carried out using *m*-CPBA in CHCl$_3$ to deliver omeprazole (**1**).

An improved transformation of **14** to **1** was patented in 1991.[7] The invention involved carrying out the *m*-CPBA oxidation of **1** in CH$_2$Cl$_2$ at a substantially higher pH of 8.0 to 8.6 using KHCO$_3$ as the buffer. The reaction mixture was subsequently extracted with aqueous NaOH. The separated aqueous phase was then treated with an alkyl formate (e.g. methylformate), maintaining the pH > 9, allowing omeprazole (**1**) to crystallize. Another improved sulfide oxidation was patented in 2000 using EtOAc as the solvent.[8]

§3.2.1 Separation using HPLC

The small-scale synthesis of esomeprazole (**2**) was conveniently accomplished via an HPLC separation of the omeprazole (**1**) enantiomers.[9] For large-scale separations, it can be achieved by a reverse phase HPLC of the two stereoisomers of a diastereomeric mixture of an alkylated omeprazole where the alkyl group serves as a chiral auxiliary.[10,11] In order to install the requisite chiral auxiliary, a "handle" was required for tethering. This was easily installed by treating **1** with formaldehyde, followed by chlorination with thionyl chloride to afford chloromethylbenzimidazole **15**. Subsequently, treatment of **15** with (*R*)-(–)-mandelic acid (**16**) in the presence of NaOH under phase transfer catalysis conditions gave ester **17** as a mixture of two diastereomers. The pair of diastereomers was then separated by reverse phase HPLC to render the pure diastereomer **17a**. Removal of the (*R*)-(–)-mandelic acid chiral auxiliary was achieved by a brief exposure of ester **17a** to a NaOH solution. Finally, magnesium salt formation using MgCl$_2$ then delivered esomeprazole (**2**).

§3.2.2 Asymmetric oxidation of the sulfide

The advent of asymmetric synthesis methodology has had a tremendous impact on both academia and industry. For example, the Sharpless epoxidation has become a staple in the construction of chiral building blocks. In this particular case,[12,13] asymmetric oxidation of the sulfide **14** was accomplished by employing a *catalytic* amount of chiral ligand, *D*-(−)-diethyl tartrate to fashion sulfoxide **18** in 99.99% *ee*. Conversion of **18** to the corresponding magnesium salt **2** was then easily accomplished by treatment with MgCl₂. This method was apparently more advantageous than the HPLC separation method which involves the use of one equivalent of chiral auxiliary. The asymmetric oxidation approach is also more amenable to the large-scale process.

§3.2.3 Biooxidation

Oxidation of the sulfide **14** can also be carried out *via* biooxidation,[14] as biotransformations have become recognized as a viable method for chemical transformations due to their environmentally friendly conditions. Thus, sulfide **14** was incubated with a variety of microorganisms in 50 mM Na_2HPO_4 buffer, pH 7.6 with 5–10 g/L dry cell weight and a substrate concentration of 1 g/L. The cells were incubated with the sulfide **14** on a rotary shaker at 28°C for 18–20 h. The screening process gave enantiomerically enriched **18** with *ee* from 56–99%. In particular, *Penicillium frequentans* BPFC 386, *Penicillium frequentans* BPFC 585 and *Brevibacterium paraffinophagus* ATCC 21494 all resulted in > 99% *ee*.

Similarly, sulfides **19** and **21** were biooxidized to the corresponding enantiomerically enriched sulfoxides, (–)-pantoprazole **20** and (–)-lansoprazole **22**, respectively. These biotransformation processes are still at the exploratory stage. The concentration of the substrate is generally minute — in the ppm range. As a result, this method is not suitable for large-scale processes.

§3.3 References

1. Lindberg, P.; Bränström, A.; Wallmark, B.; Mattsson, H.; Rikner, L.; Hoffman, K.-J. *Med. Res. Rev.* **1990**, *10*, 1–54.

2. Lindberg, P.; Nordberg, P.; Almınger, T.; Bränström, A.; Wallmark, B. *J. Med. Chem.* **1986**, *29*, 1327–1329.

3. Junggren, U. K.; Sjöstrand, S. E. EP 0005129 (**1979**).

4. Kohl, B.; Sturn, E.; Senn-Bilfinger, J.; *et al. J. Med. Chem.* **1992**, *35*, 1049–1053.

5. Winterfeld, K.; Flick, K. *Arch. Pharm.* **1956**, *26*, 448–452.

6. Li, J. J. *Name Reactions* Springer: Berlin, **2002**, 39.

7. Brändströn, A. E. WO 9118895 (**1991**).

8. Hafner, M.; Jereb, D. WO 0002876 (**2000**).

9. Erlandsson, P.; Isaksson, R.; Lorentzon, P.; Lindberg, P. *J. Chromatography* **1990**, *532*, 305–319.

10. WO 9208716 (**1992**).

11. Lindberg, P. L.; Von Unge, S. WO 9427988 (**1994**).

12. Lindberg, P. L.; Von Unge, S. US 5714504 (**1998**).

13. Cotton, H.; Kroström, A.; Mattson, A.; Möller, E. WO 9854171 (**1998**).

14. Holte, R.; Lindberg, P.; Reeve, C.; Taylor, S. WO 9617076 (**1996**).

Chapter 4. Protein-tyrosine Kinase Inhibitors:
Imatinib (Gleevec®) and Gefitinib (Iressa®)

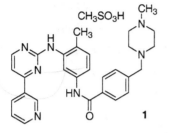

USAN. Imatinib Mesylate
Trade Name: Gleevec®
Novartis
Launched: 2001
M W 493.6 (parent)

USAN: Gefitinib
Trade Name: Iressa®
AstraZeneca
Launched: 2003
M.W. 446.9

2

§4.1 Introduction to Gleevec®[1-5]

Chronic myeloid leukemia (CML) is a rare but life-threatening cancer related to marrow. It is a hematologic stem cell disorder caused by an acquired abnormality in the DNA of the stem cells in bone marrow. One of the treatment options is bone marrow transplant. Not only is a matching donor needed, but the procedure is highly risky. Other treatments of CML involve chemotherapy such as Ara-C (an intravenous drug), hydroxyurea (oral chemotherapy), and interferon (administered by daily injection). Gleevec® represents a new paradigm in cancer therapy. It revolutionized the treatment of CML and gastrointestinal stromal tumors (GIST) patients as an oral drug with relatively fewer and more tolerable side effects in comparison to Ara-C, hydroxyurea, and interferon. Gleevec® (imatinib mesylate, **1**) is a protein kinase inhibitor, the first marketed drug whose mechanism of action (MOA) is through inhibition of a protein kinase, functioning as a signal transduction inhibitor (STI). The FDA of the United States approved this drug in a record time of 2.5 months for the treatment of CML in May 2001. In 2002, it received approval from the FDA as the first-line treatment of CML and for the indication of GIST.

    Protein kinase inhibitors belong to a class of chemotherapy that disrupts the signal transduction system within the cell. On the contrary, protein kinases modulate intracellular signal transduction by catalyzing the phosphorylation of specific proteins. In contrast, protein phosphatases function through dephosphorylation of proteins to modulate biological activities. Many cellular processes are the results of the interplay between phosphorylation by protein kinases and dephosphorylation by protein phosphatases. Therefore, protein kinase inhibition is an area for therapeutic intervention against a variety of diseases such as cancer, inflammatory disorders and diabetes. It is worth noting that although there are over 5000 protein kinases in human, there are but only a few phosphatases. Interestingly, most kinase inhibitors are flat aromatic compounds that mimic the adenine portion of ATP (3).

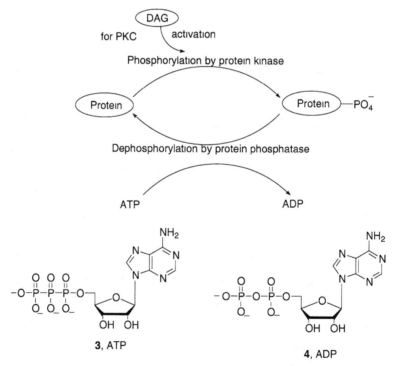

Fig. 1. The function of protein kinases [ADP (4), adenosine diphosphate; ATP (3), adenosine triphosphate].

Protein kinases can be classified according to the amino acid residue that is phosphorylated in the cellular process. Consequently, there are tyrosine-specific kinases and serine/threonine kinases. Tyrosine kinases are a family of tightly regulated enzymes, and the aberrant activation of various members of this family is one of the hallmarks of cancer. Tyrosine phosphorylation has been linked to multiple cell growth and differentiation pathways. Imatinib mesylate (1) is a tyrosine kinase inhibitor (TKI). An important characteristic of imatinib mesylate (1) is that it is an *ATP-competitive* inhibitor. It binds at the ATP binding site and blocks ATP binding thereby inhibiting kinase activities.

Phenylaminopyrimidine 5 was initially identified from a random screening as a PKC-α inhibitor with an IC_{50} of approximately 1 μM. It did not inhibit Abl (an abnormal protein tyrosine kinase) and PDGFR (platelet-derived growth factor receptor, a kinase). Intensive SAR studies involving more than 300 analogs eventually led to imatinib mesylate (1). Significantly, the introduction of the "flag-methyl group" at the 6-position of the phenyl ring (highlighted in Figure 2) abolished the activity against PKC-α. In the end, 1 was obtained as a selective Bcr-Abl-tyrosine kinase inhibitor. Bcr-Abl is an abnormal protein tyrosine kinase produced by the specific chromosomal abnormality called the Philadelphia chromosome. This chromosome is a marker for chronic myeloid leukemia (CML). In addition, the drug also inhibits a third tyrosine kinase receptor, the c-kit receptor that is associated with GIST.

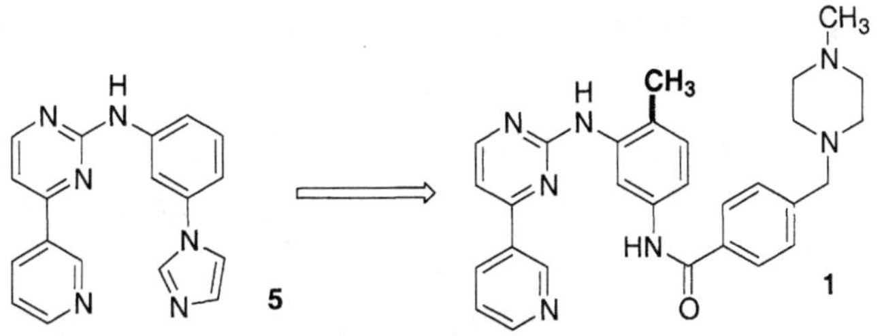

Fig. 2. Evolution of the screening lead 5 to imatinib mesylate (1).

The enzymatic *in vitro* test of 1 against v-Abl-tyrosine kinase showed that it is a potent inhibitor with an IC_{50} = 38 nM. However, in cell, the IC_{50} is only 250 nM for the inhibition of the autophosphorylation of v-Abl. Nonetheless, what makes **1** an efficient drug is its excellent bioavailability (98% in human), which allows for once daily oral treatment. Additional pharmacokinetics parameters of imatinib mesylate (**1**) for human are listed in Table 1.

	C_{max}	T_{max}	$T_{1/2}$	AUC_{0-24}
	µg/L	hr	hr	µg*hr /L
Imatinib mesylate (**1**) 600 mg p. o.	3925	0.5–3.0	10–23	59535

Table 1. Pharmacokinetics parameters of imatinib mesylate (**1**).

§4.2   Synthesis of imatinib mesylate[6–11]

The 12-step synthesis of imatinib mesylate (**1**) in the manufacturing process was accomplished by Novartis in an astonishingly short time. The synthesis began with a condensation reaction between **6** and ethyl formate. Deprotonation of the methyl group on 3-acetylpyridine (**6**) using freshly prepared sodium methoxide afforded an enolate. Condensation of the enolate with ethyl formate was followed by exchange with dimethylamine to produce 3-dimethylamino-1-(3-pyridyl)-2-propen-1-one (**7**). Alternatively, **7** could be prepared from the condensation of **6** and *N,N*-dimethylformamide dimethylacetal [(MeO)$_2$CHN(Me)$_2$].[6,7] On the contrary, nitration of 2-amino-toluene (**8**) gave 2-amino-4-nitrotoluene nitrate, with the nitro group serving as a masked amine group. Refluxing 2-amino-4-nitrotoluene nitrate with cyamide furnished 2-amino-5-nitrophenylguanidine (**9**). Subsequently, formation of pyrimidine **10** was achieved from condensation of 3-dimethylamino-enone **6** and guanidine **9** in the presence of NaOH in refluxing isopropanol.[8,9] Palladium-catalyzed hydrogenation in THF of nitrophenylpyrimidine **10** unmasked the nitro group to provide aminophenylpyrimidine **11**. Amide formation was accomplished by treatment of aniline **11** with 4-(4-methypiperazinomethyl)-benzoyl chloride (**12**) to deliver **1**. Finally, the mesylate of **1** was readily accessed by the addition of one equivalent of methanesulfonic acid.[10,11]

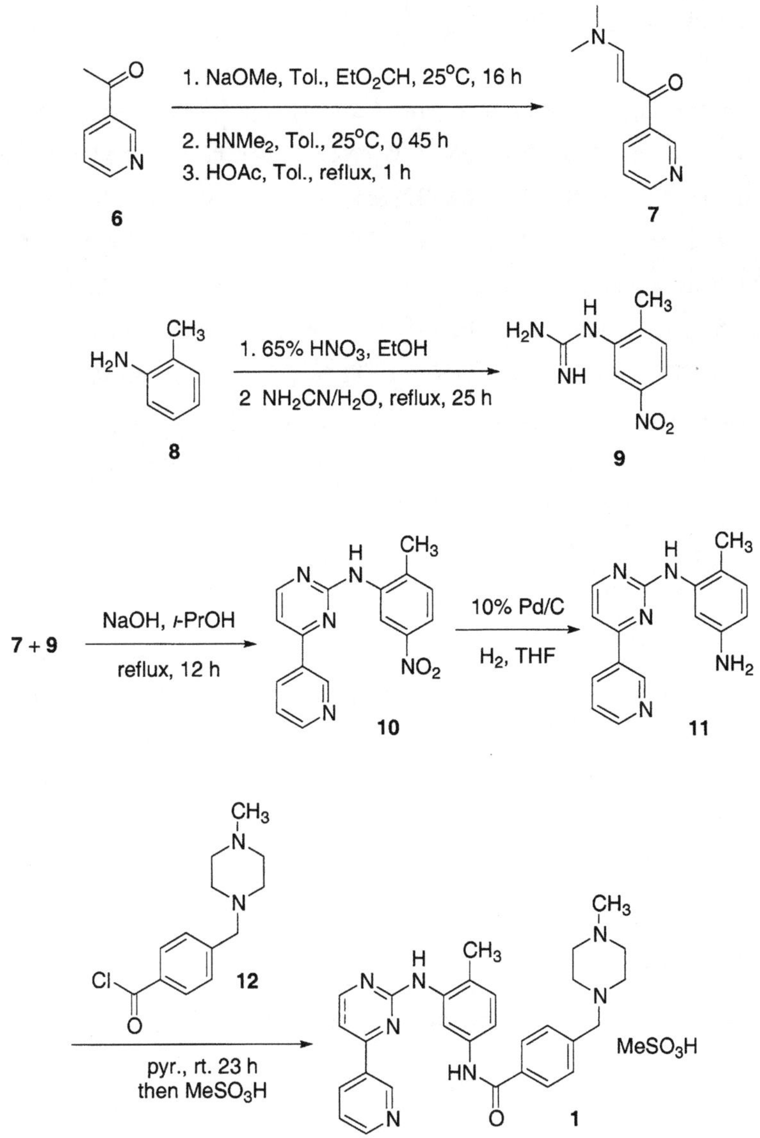

Scheme 1. Synthesis of imatinib mesylate (1).

§4.3   Introduction to Iressa[®12–19]

Epidermal growth factor is a tyrosine kinase. Epidermal growth factor receptor (EGFR) is a member of the *erb*B family. Genetech's monoclonal antibody, trastuzumab (Herceptin[®]) targets the Her-2/neu (*erb*B2) receptor. Her-2 stands for **h**uman **e**pidermal growth factor **r**eceptor-**2**, whereas neu is the short form of **neu**rological tumors. Herceptin[®], representing a revolutionary treatment for metastatic breast cancer, was approved by the FDA in 1989. Iressa[®] (gefitinib, **2**) is an epidermal growth factor receptor (EGFR) tyrosine kinase inhibitor (TKI). It works by disrupting molecular signals that turn normal cells into tumors. As such, it does not damage healthy cells or cause side effects as harsh as those associated with standard chemotherapy treatments. Iressa[®] (**2**) won FDA approval early in 2003.

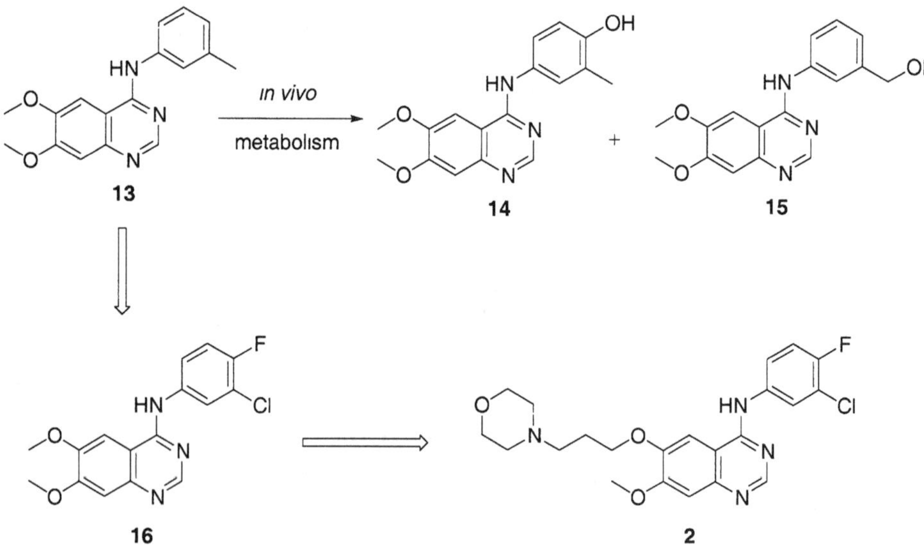

Fig. 3. Evolution of **13** to Iressa[®] (**2**).

4-Anilinoquinazoline **13** was one of the older EGFR-TKI. Despite its potency (IC_{50} = 5 nM), its half-life in mice is only 1 h. The major metabolizing pathways are oxidation of **13** to phenol **14** and benzyl alcohol **15**. Blocking the two metabolizing sites afforded 4-anilinoquinazoline **16**, which has a half-life of 3 h in mice.   Upon further optimization of the molecule with a basic 6-alkyl side chain, gefitinib (**2**) was obtained with improvement of *in vivo* activities as well as physical properties, which in turn afforded a favorable pharmacokinetics attribute that allows a once-daily oral dosing.

Iressa® (gefitinib, **2**) is a reversible inhibitor, thus mechanistically it is possible to excert less toxicity in comparison to similar irreversible inhibitors, which bond to the protein covalently.   Gefitinib (**2**) is also quite selective.   It is at least 100-fold selective against other tyrosine kinases such as *erb*B-2, KDR, c-flt or serine/threonine kinases such as PKC, MEK-1, and ERK-2.

§4.4   Synthesis of gefitinib[20,21]

The AstraZeneca synthesis of Iressa® (gefitinib, **2**) began with selective mono-demethylation of 6,7-dimethoxy-3H-quinazolin-4-one (**17**) using *L*-methionine (**18**) in methanesulfonic acid to afford phenol **19**.   Acetylation of **19** gave acetate **20** in 75% yield in 2 steps.   Chlorination of the quinazolin-4-one using thionyl chloride with the aid of a catalytic amount of DMF provided chloro-quinazoline **21**.   The subsequent S_NAr displacement of **21** with 3-chloro-4-fluoroaniline (**22**) assembled **23**.   Deprotection of the acetate using methanolic ammonia gave the key intermediate, phenol **24**.   Finally, alkylation of **24** with alkyl chloride **25** using K_2CO_3 in DMF then delivered gefitinib (**2**).

In conclusion, the conquest of cancer treatment in human history has come a long way. In the 1950s, mustard gas was found to alkylate white blood cells preferentially, heralding the modern era of chemotherapy. Since then, 5-FU, taxol, methoxotrexate, temoxefen, and many more chemotherapeutic drugs have been discovered, enriching the arsenal of cancer treatment. Unfortunately, they destroy not only the cancer cells, but damage the normal cells as well, imparting toxic side effects. Gleevec® and Iressa® as small molecular drugs, along with Herceptin® as a biologic, are so-called "smart drugs". They target only the oncogenes such as Bcr-Abl and Her-2/neu without interrupting the healthy cells. They marked the beginning of a very exciting age of discovery of "magic bullet" cancer drugs.

§4.3   References

1.   Traxler, P.; Bold, G.; Buchdunger, E.; Caravattı, G.; Furet, P.; Manley, P.; O'Reilly, T.; Wood, J.; Zimmermann, J. *Med. Res. Rev.* **2001**, *21*, 499–512.

2.   Dumas, J. *Expert Opinion on Therapeutic Patents* **2001**, *11*, 405–429.

3.   de Bree, F.; Sorbera, L. A.; Fernández, R.; Castañer, J. *Drugs Fut.* **2001**, *26*, 545–552.

4.   Kumar, C. C.; Madison, V. *Expert Opinion on Emerging Drugs* **2001**, *6*, 1765–1774.

5.   Lyseng-Wıllıamson, K.; Jarvis, B. *Drugs* **2001**, *61*, 303–315.

6.   Zimmermann, J.; Buchdunger, E.; Mett, H.; Meyer, T.; Lydon, N. B. *Bioorg. Med. Chem. Lett.* **1997**, *7*, 187–192.

7.   Zimmermann, J. US 5521184 (**1996**).

8.   Torley, L. W.; Johnson, B. B.; Dusza, J. P. EP 0233461 (**1987**).

9.   Dusza, J. P.; Albright, J. D. US 4281000 (**1981**).

10. Druker, B. J.; Talpaz, M.; Resta, D. J.; Peng, B.; Buchdunger, E.; Ford, J. M.; Lydon, N. B.; Kantarjian, H.; Capdeville, R.; Ohno-Jones, S.; Sawyers, C. L. *N. Engl. J. Med.* **2001**, *344*, 1031.

11. Druker, B. J.; Sawyers, C. L.; Kantarjian, H.; Resta, D. J.; Reese, S. F.; Ford, J. M.; Capdeville, R.; Talpaz, M. *N. Engl. J. Med.* **2001**, *344*, 1038–1042.

12. *Iressa*, Cıardiello, F. *Drugs,* **2000**, *60*, 25–32.

13. Baselga, J.; Averbuch, S. D. *Drugs,* **2000**, *60*, 33–40.

14. Adjei, A. A. *Drugs Fut.* **2001**, *26*, 1087–1092.

15. Levin, M.; D'Souza, N.; Castañer, J. *Drugs Fut.* **2002**, *27*, 339–345.

16. Ranson, M.; Mansoor, W.; Jayson, G. *Expert Review of Anticancer Therapy* **2002**, *2*, 161–168.

17. Culy, C. R; Faulds, D. *Drugs* **2002**, *62*, 2249–50.

18. Herbst, R. S. *Expert Opinion on Investigational Drugs* **2002**, *11*, 837–849.

19. Laird, A. D.; Cherrington, J. M. *Expert Opinion on Investigational Drugs* **2003**, *12*, 51–64.

20.   Gibson, K. H. WO 96/33980 (**1996**).

21.   Barker, A. J.; Gibson, K. H.; Grundy, W.; Godfrey, A. A.; Barlow, J. J.; Healy,
      M. P.; Woodburn, J. R.; Ashton, S. E.; Curry, B. J.; Scarlett, L.; Henthorn, L.;
      Richards, L. *Bioorg. Med. Chem. Lett.* **2001**, *11*, 1911–1914.

Chapter 5. Non-sedating Antihistamines

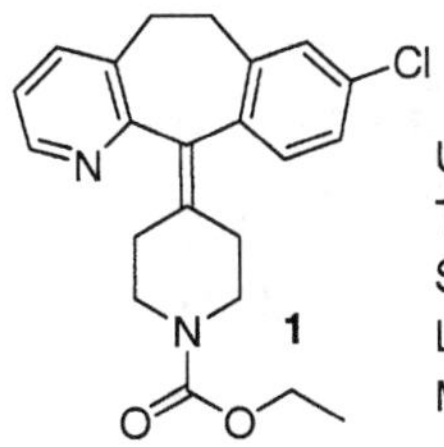

USAN: Loratadine
Trade Name: Claritin®
Schering–Plough Corp.
Launched 1989
M.W. 382 9

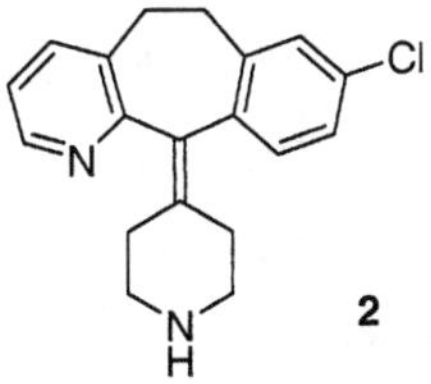

1

USAN: Desloratadine
Trade Name· Clarinex®
Schering–Plough Corp.
Launched· 2002
M.W. 310.8

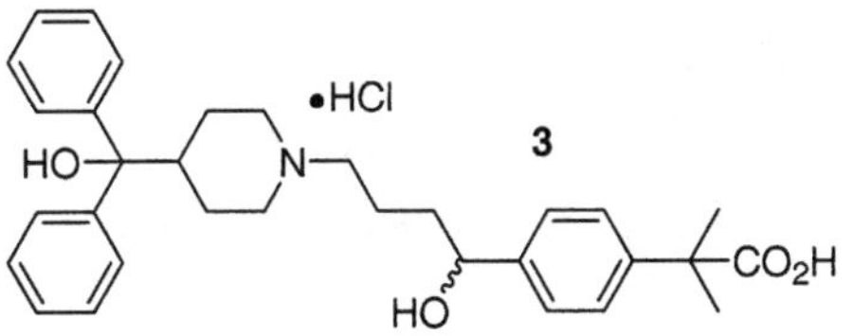

2

•HCl

3

USAN: Fexofenadine
Trade Name: Allegra®
Aventis
Launched. 1996
M.W 501.7 (parent)

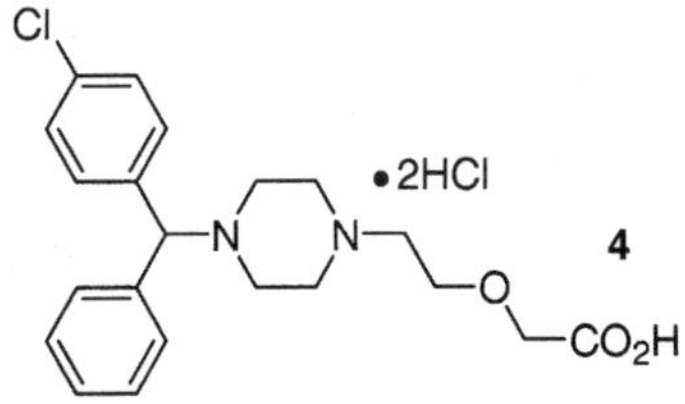

•2HCl

4

USAN Cetirizine dihydrochloride
Trade Name Zyrtec®
Pfizer Inc./UCB S. A.
Launched: 1996
M.W. 388.9 (parent)

§5.1 Introduction[1–3]

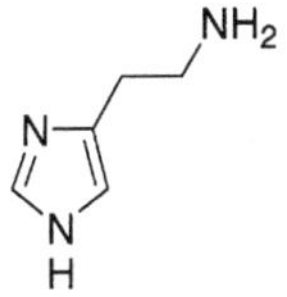

5, histamine

Antihistamines are antagonists of histamine receptors that displace histamine competitively from its receptors and block the effects of histamine. Histamine receptors, in turn, are G-protein-coupled receptors (GPCRs) that contain the typical seven-transmembrane loop motif. Other common GPCRs include calcium channel receptors, andrenergic α_1, dopamine D_2, serotonin $5\text{-}HT_2$ and muscarinic receptors. They have traditionally been productive targets for drug discovery. Histamine receptor inhibitors alone include several blockbuster drugs such as famotidine (for ulcers), ranitidine (for ulcers), loratadine (for allergies), fexofenadine (for allergies), and cetirizine (for allegies). Moreover, many drugs such as fenoldopam, captopril, prazosin that serve as dopamine, serotonin, angiotensin and adrenergic receptor antagonists are marketed. Indeed, nowadays, GPCR inhibitors account for 20% of the top 50 best selling drugs and greater than 50% of all drugs marketed today.

The use of antihistamines can be traced back to the beginning of 1940s. The applications of the first-generation antihistamines were limited since they cause significant adverse effects such as sedation, memory impairment and psychomotor dysfunction. The second-generation antihistamines have significantly fewer central nervous system (CNS) adverse effects because they penetrate the blood-brain barrier much less extensively.

§5.2   Synthesis of loratadine (1)[4-6] and desloratadine (2)[7-10]

The evolution of azatadine (6) to loratadine (1) and desloratadine (2) is an excellent example that shows that a minor change in the molecular structure can have an enormous impact on the pharmacological profile. In 1973, Schering–Plough launched their first-generation antihistamine azatadine (6, Optimine®). Like all first-generation antihistamines, azatadine (6) was found to penetrate the blood-brain barrier and cause severe drowsiness. Simple treatment of 6 with ethyl chloroformate results in dechloro-loratadine (7), which is nearly devoid of the sedating CNS effects. However, the systemic clearance of 7 takes place rapidly, with an effective duration of only 6–8 hrs. On the other hand, addition of the 8-chloro-substituent gave loratadine (1), which provides at least 50% protection for up to 18 h when tested at 0.25 and 0.5 mg/Kg doses. Thus, a once daily regimen is amenable. Furthermore, loratadine (1) is four times more potent than dechloro-loratadine 7. In addition, loratadine (1) displays little activity at acetylcholine or adrenergic receptors and is inactive in animal models for assessing anticholinergic effects at doses up to 320 mg/Kg.

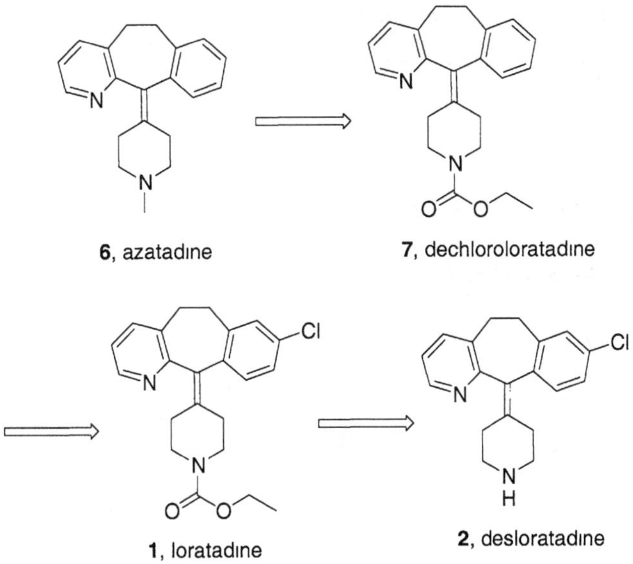

Fig. 1. Evolution of azatadine (6) to loratadine (1) and desloratadine (2).

The synthesis of azatadine (6) was patented in 1967. Although the initial discovery syntheses of loratadine (1) and desloratadine (2) were lengthy, the process chemists came up with an elegant synthesis as shown in Scheme 1.[4-6] At first, the Ritter reaction of 2-cyano-3-methylpyridine (10) with tert-butanol was carried out at 75 °C with the aid of concentrated sulfuric acid to afford tert-butyl carboxamide 11 (the interesting mechanism of the Ritter reaction involves a nucleophilic attack of the nitrogen atom on the nitrile group to the tert-butyl cation). Subsequent deprotonation of the methyl group on 11 using two equivalents of n-BuLi was followed by addition of catalytic amount of sodium bromide and then m-chlorobenzyl chloride to give adduct 12. The catalytic sodium bromide underwent a Finklestein reaction with m-chlorobenzyl chloride to afford m-chlorobenzyl bromide, which is a better electrophile for the S_N2 reaction. Refluxing 12 with POCl$_3$ for 3 h transformed the tert-butyl carboxamide back to the nitrile group to give 13. Addition of N-methyl-piperidyl magnesium chloride (14) to the nitrile group of 13 gave the corresponding imine magnesium bromide, which was hydrolyzed to deliver ketone 15. Cyclization of 15 using an excess super acid composed of HF and BF$_3$ led to cycloheptene 16 in 92% yield. Concurrent demethylation and carbamate 1 formation took place when **16** was treated with ethyl chloroformate at 80°C. Hydrolysis of loratadine (**1**) was then carried out using KOH in refluxing $H_2O/EtOH$ to deliver desloratadine (**2**).

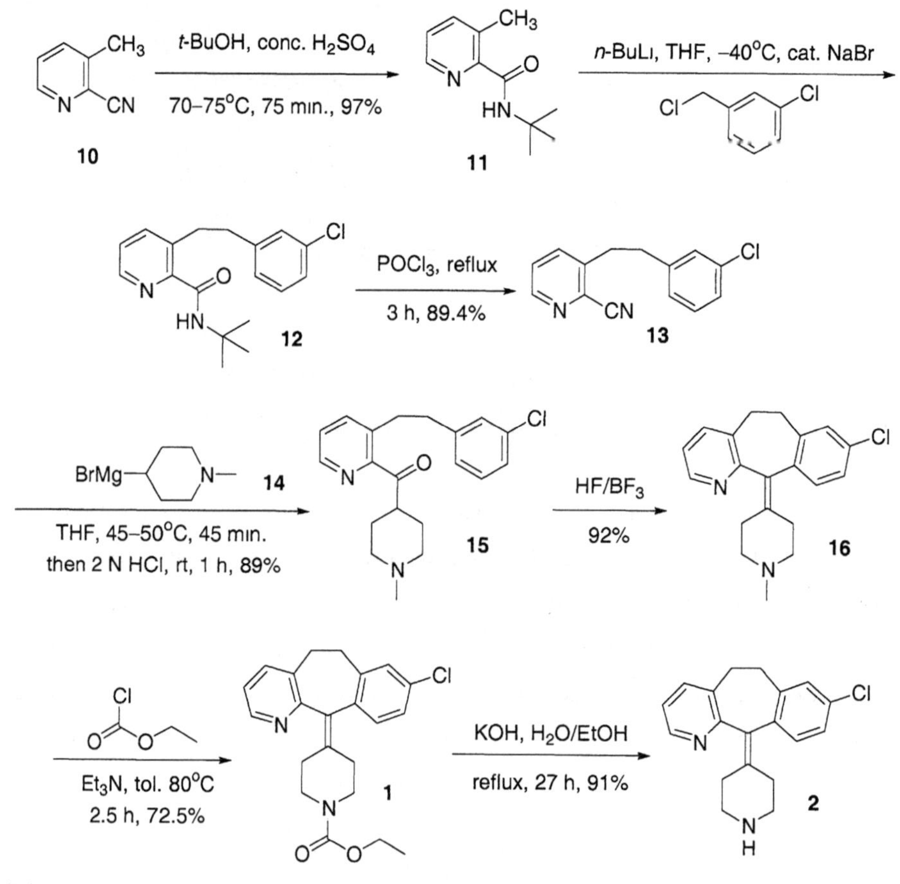

Scheme 1. Synthesis of loratadine (**1**) and desloratadine (**2**).

Alternatively, as shown in Scheme 2, the von Braun reaction of **16** using cyanogen bromide produced the *N*-cyanopiperidine **17**. Acidic hydrolysis of **17** in refluxing glacial acetic acid containing concentrated HCl then provided desloratadine (**2**) in 93% yield.

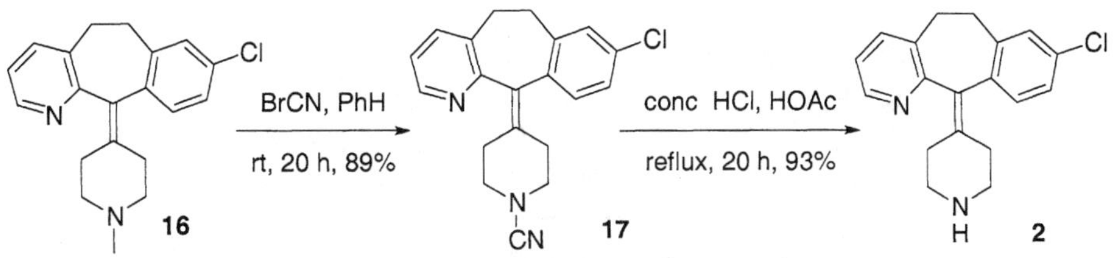

Scheme 2. An alternative synthesis of desloratadine (2).

§5.3 Synthesis of fexofenadine (3)[11-19]

In 1994, terfenadine (18, Seldane®) was the world's leading non-sedating antihistamine. However, it was found to prolong the QT_c interval in the ECG which was indicative of serious ventricular arrhythmia, torsades de pointes, especially when it was taken concomitantly with drugs known to alter hepatic oxidative metabolism such as erythromycin, ketoconazole and macrolide antibiotics. The drug-drug interactions stem from the fact that these drugs are all metabolized by the same cytochrome P450 isoenzyme, CYP3A4. On the contrary, fexofenadine (3), the major metabolite of terfenadine (18), is devoid of the QT_c elongation problem and has less propensity to cause drug-drug interactions. Therefore, terfenadine (18) can be considered a pro-drug of fexofenadine (3). It is advantageous to administer fexofenadine (3) rather than terfenadine (18) because the liver does not have to be burdened with the metabolism process of 18. In a rare move, the FDA worked closely with the manufacturer to replace terfenadine (18) with fexofenadine (3).

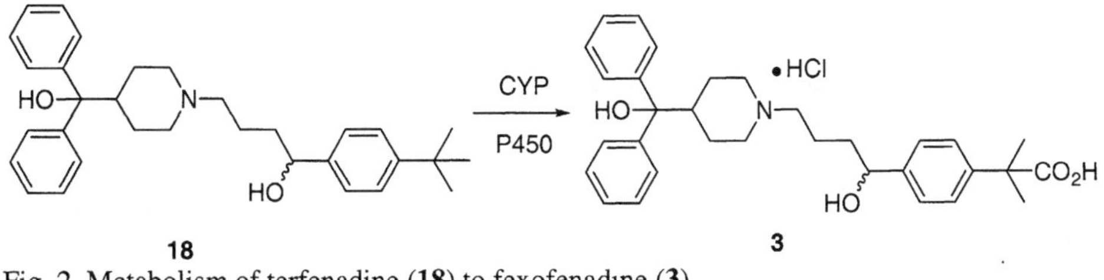

Fig. 2. Metabolism of terfenadine (18) to fexofenadine (3).

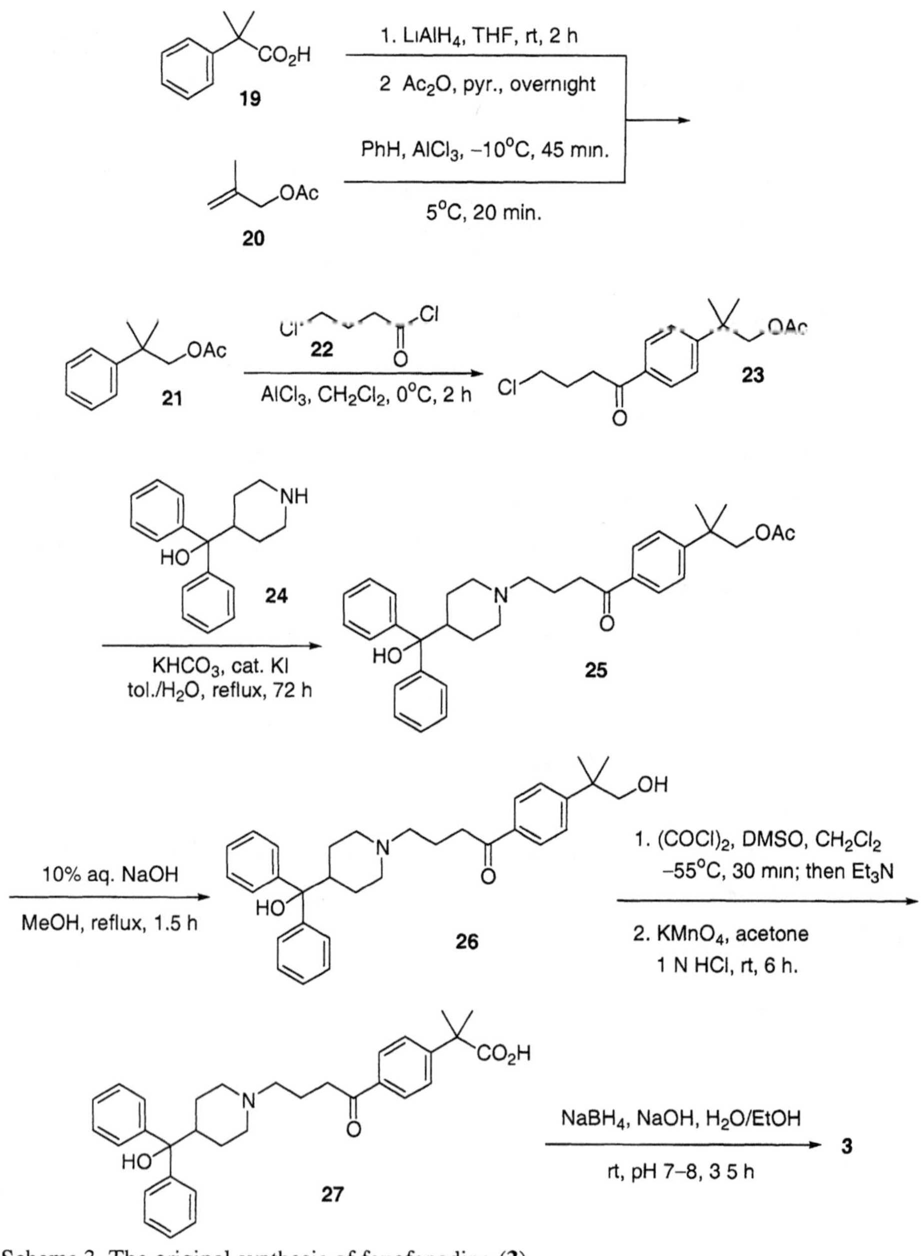

Scheme 3. The original synthesis of fexofenadine (3).

The synthesis of fexofenadine (**3**) by Merrell Dow commenced by preparation of α,α-dimethylphenethyl acetate (**21**).[14] Reduction of α,α-dimethylbenzyl acetic acid (**19**) led to α,α-dimethylbenzyl alcohol, which was subsequently acetylated to **21** using acetic anhydride. Alternatively, **21** could be assembled by a Friedel–Crafts reaction using benzene and 2-methyl-2-propenyl acetate (**20**) in one step. An additional Friedel–Crafts reaction of **21** with ω-chlorobutyryl chloride (**22**) was mediated by one equivalent of AlCl$_3$ to produce adduct **23**. The S$_N$2 displacement of the chloride on **23** by α,α-diphenethyl-4-piperidinemethanol (**24**, *vide infra*) was aided by addition of a catalytic amount of KI *via* the Finkelstein reaction mechanism to give **25**. Removal of the acetate protection was easily achieved using basic hydrolysis to afford neopentyl alcohol **26**. Oxidation of **26** using the Swern oxidation followed by oxidation with KMnO$_4$ gave the corresponding carboxylic acid **27**. The transformation **26** → **27** was also accomplished using either H$_5$IO$_6$/RuCl$_3$•5H$_2$O in CHCl$_3$/CH$_3$CN or K$_2$S$_2$O$_8$/RuCl$_3$•5H$_2$O. Finally, reduction of ketone **27** then delivered fexofenadine (**3**). It is worth mentioning that asymmetric reduction of **27** was also carried out using either (+) or (−)-*B*-chloro-diisopinocamphenylborane, leading to the (*R*)- and (*S*)-isomers, respectively.

In 1994, the Just group at McGill University published their version of fexofenadine (**3**) synthesis.[15] Methyl esterification of 4-bromophenylacetic acid (**28**) using 2.2 equivalents of Me$_3$SiCl in methanol was followed by double methylation of the benzylic carbon using 2.2 equivalents of MeI using NaH in THF to give **29**. The crucial convergency was achieved by the Sonogashira reaction of **29** and 3-butyne-1-ol catalyzed by (Ph$_3$P)$_4$Pd/CuBr$_2$ in refluxing Et$_3$N to produce **30**. A simple mesylation of **30** led to mesylate **31**, which was subjected to an S$_N$2 reaction with piperidine **24** to assemble adduct **32**. The mercury-catalyzed hydration of alkyne **24** afforded benzylic ketone **33**, which was reduced by NaBH$_4$ to deliver fexofenadine (**3**) after hydrolysis of the methyl ester.

The aforementioned synthetic route is evidently efficient. However, the use of mercury(II) oxide precludes it from being used as an industrial production route.

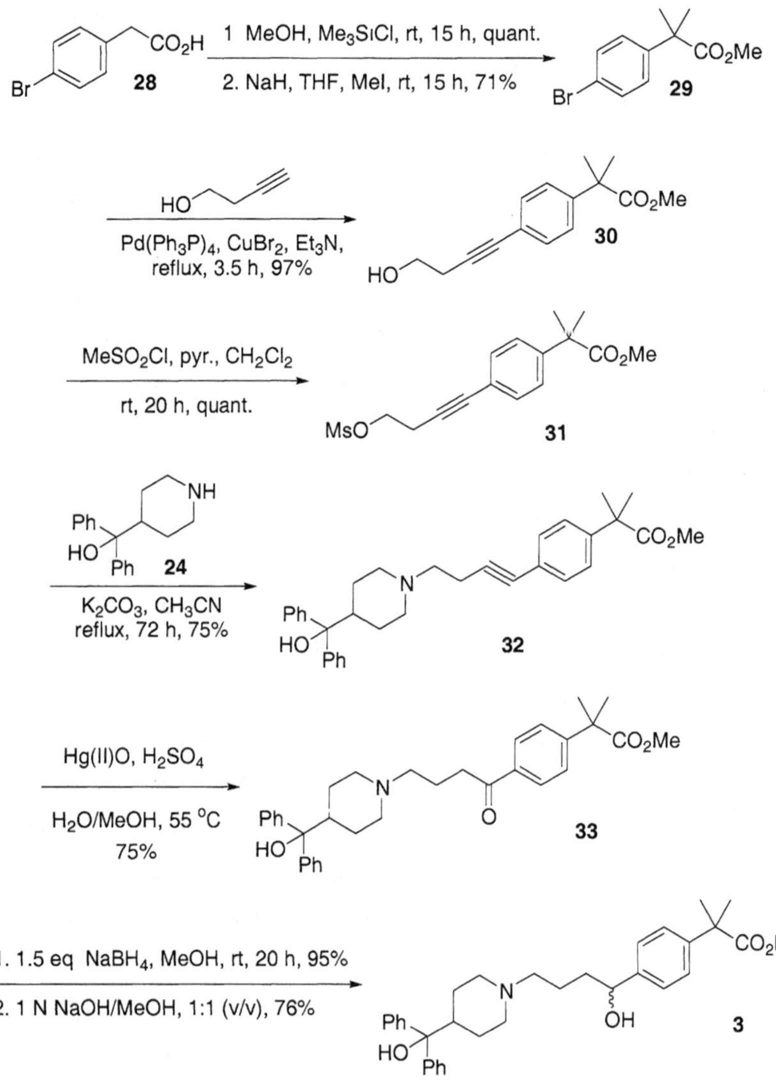

Scheme 4. The McGill synthesis of fexofenadine (3) involving a Sonogashira reaction.

In 1996, a group from Sandoz (now part of Novartis) reported their synthesis of fexofenadine (3).[17] Cbz protection of commercially available ethyl piperidine-4-carboxylate (34) was achieved using N-(benzyloxycarbonyloxy)succinimide to afford 35. Treatment of 35 with three equivalents phenyl magnesium bromide led to tertiary alcohol

36, which was then unmasked to give diphenylpiperidin-4-ylmethanol (24) via hydrogenolysis. Alkylation of 24 with alkyl chloride 37 was facilitated by KI *via* the Finkelstein reaction to produce dioxolane 38 which was unmasked to aldehyde 39 *via* acidic hydrolysis. Finally, addition of the organolithium reagent, derived from phenyl bromide 40 using one equivalent of NaH and two equivalent of *t*-BuLi, to aldehyde 39 delivered fexofenadine (3).

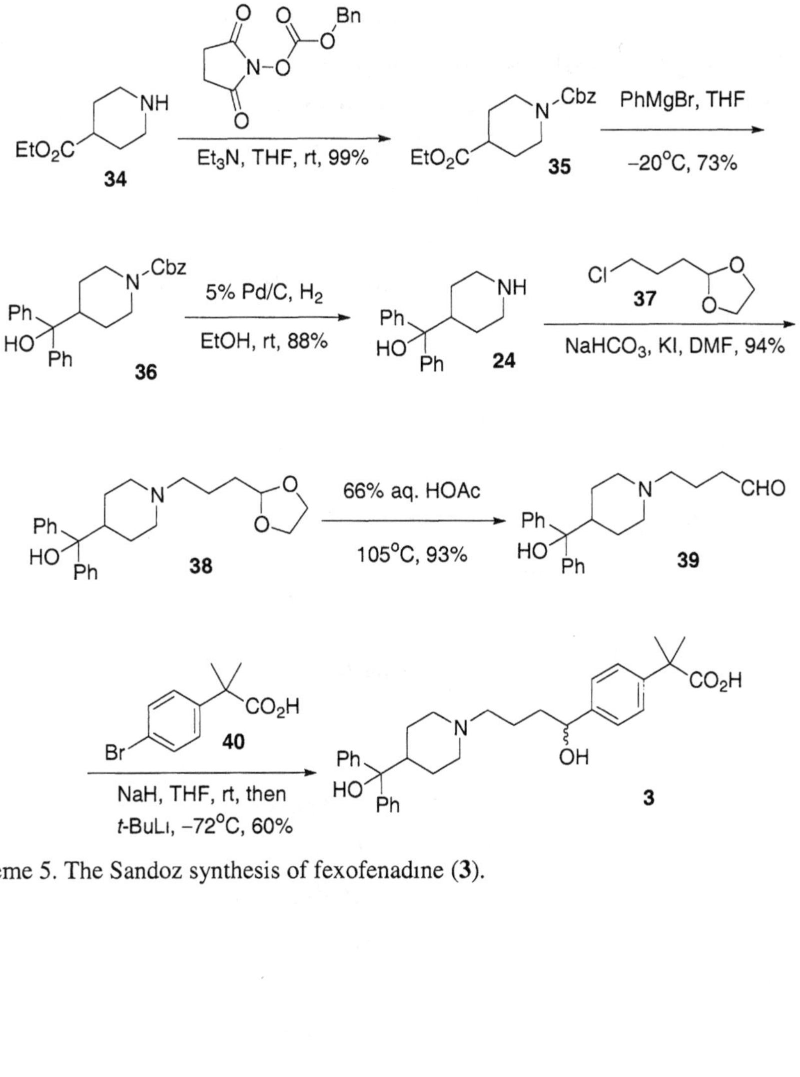

Scheme 5. The Sandoz synthesis of fexofenadine (3).

§5.4   Synthesis of cetirizine (4)[20–24]

Hydroxyzine (**41**, Fig. 3) has been used as a diphenylmethane-type of tranquilizer since the 1960s. In humans, hydroxyzine (**41**) is metabolized into cetirizine (**4**, 2-[2-[[4-[(4-chlorophenyl)phenylmethyl]-1]piperazinyl]ethoxy]-acetic acid) via an oxidation process mediated by the CYP3A4 enzyme system. Cetirizine (**4**) is a specific and long-acting histamine H_1-receptor antagonist, that is associated with a significantly lower incidence of sedation than hydroxyzine (**41**). Looking at the structures of **4** versus **41**, the only difference is a carboxylic acid versus an alcohol. Therefore, it is reasonable to speculate that the increased polarity of cetirizine (**4**) may be responsible for lowering the blood-brain barrier penetration. Another attribute that contributes to fewer CNS effects for cetirizine (**4**) is that it is more selective for histamine H_1-receptors in comparison with other nonhistamine G-protein-coupled receptors (GPCRs) including calcium channel receptors, andrenergic α_1, dopamine D_2, serotonin 5-HT_2 receptors and muscarinic receptors.[20,21]

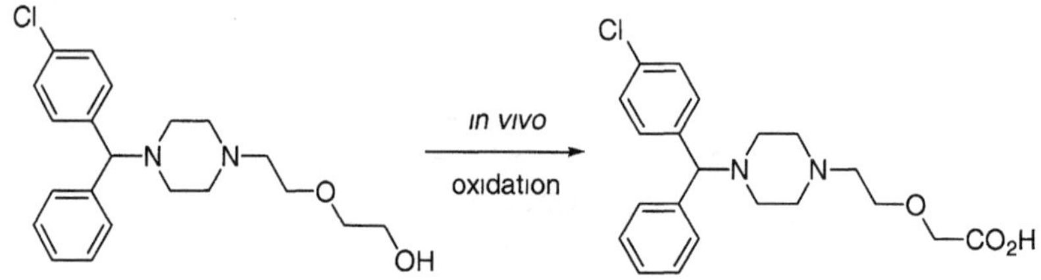

Fig. 3. Evolution of hydroxyzine (**41**) to cetirizine (**4**).

In term of pharmacokinetics, cetirizine (**4**) is rapidly absorbed, reaching a peak plasma concentration of 257 μg/L within 1 h of administration at a dose of 10 mg in healthy volunteers. The AUC is 2.87 mg/L with this dose.

The initially patented synthesis of cetirizine (**4**) by UBC S. A. in Belgium was an extension of the hydroxyzine (**41**) synthesis.[22–25] Grignard addition to *p*-chlorobenzaldehyde gave (4-chloro-phenyl)-phenyl-methanol, which was easily brominated using PBr$_3$ in benzene to give 4-chlorodiphenylmethyl bromide (**42**).[22] S_N2

displacement of the bromide using *N*-ethoxycarbonylpiperizine was carried out in the presence of Na_2CO_3 in refluxing xylene. Hydrolysis using KOH in EtOH afforded piperizine **43**.[23] An additional S_N2 displacement of 2-chloroethanol by **43** then gave alcohol **44**, which was subsequently coupled with sodium chloroacetate, followed by bis-HCl salt formation and recrystallization with 2-butanone to deliver cetirizine dihydrochloride (**4**).[24] Alternatively, **4** could be assembled from **43** and the entire side-chain [(2-chloroethoxy)-acetic acid derivative] as one fragment.[25]

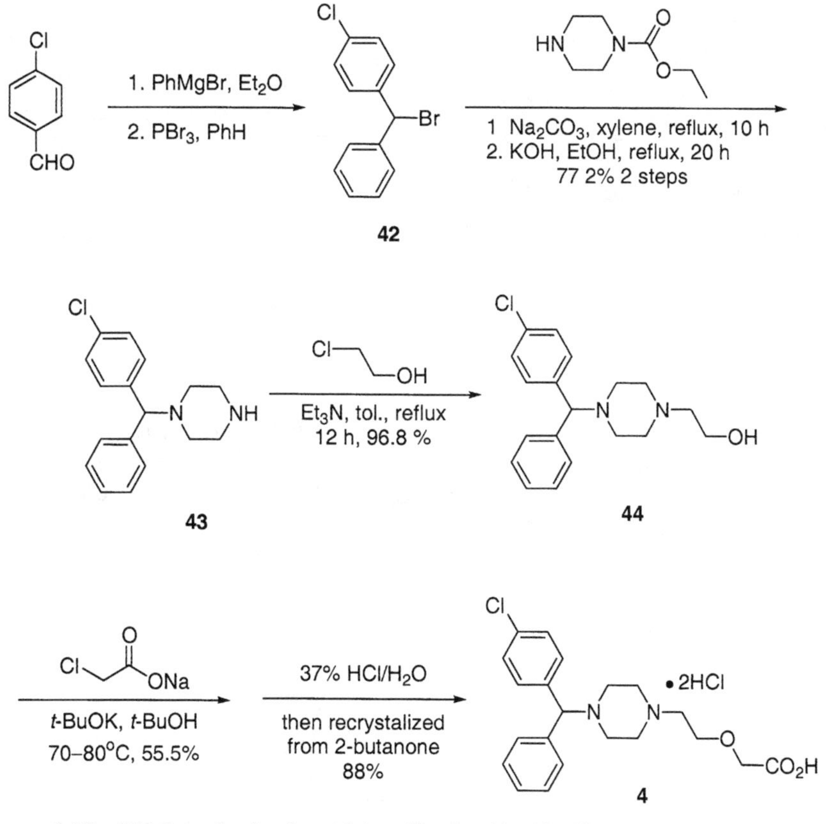

Scheme 6. The UBC synthesis of cetirizine dihydrochloride (**4**).

Since cetirizine dihydrochloride (**4**) is a mixture of two enantiomers, they have been separated and tested individually. The levorotatory enantiomer of cetirizine displays a better pharmacological profile than the racemic mixture, and is currently marketed as Xyzal[(TM)] in Europe. Meanwhile, several asymmetric syntheses to the levorotatory enantiomer have appeared in the literature.[26–29]

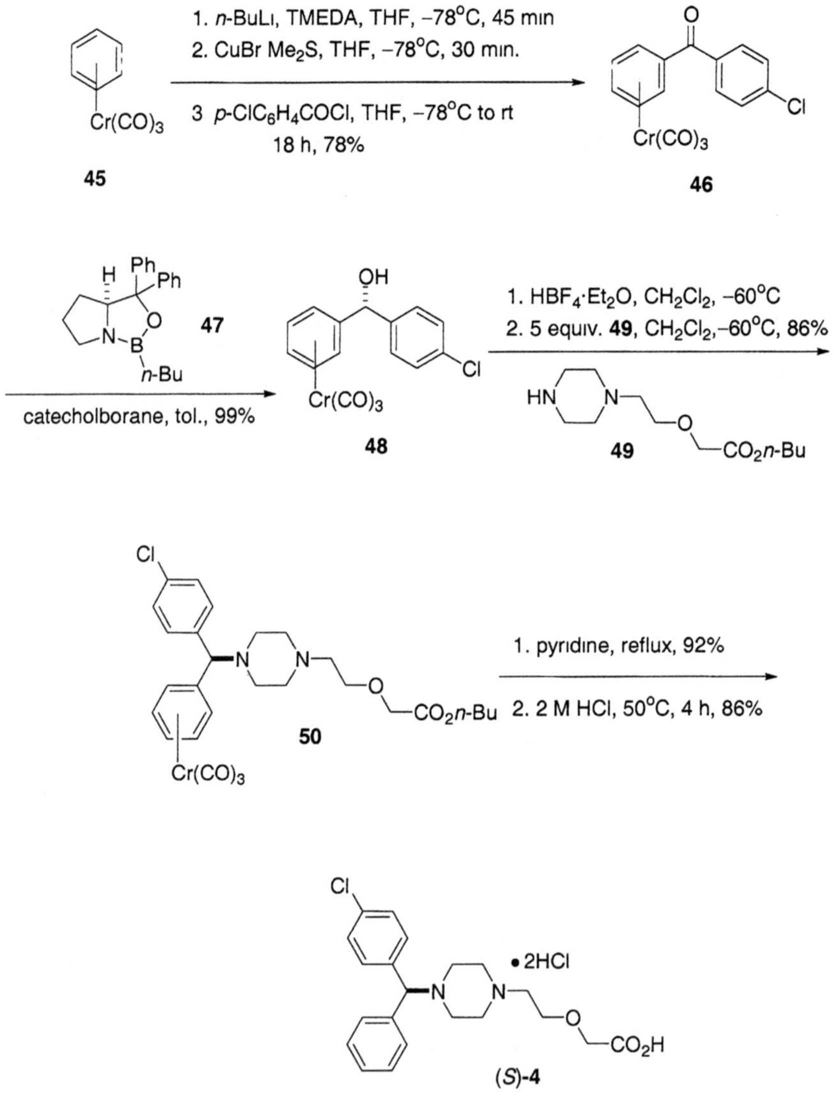

45

1. *n*-BuLi, TMEDA, THF, –78°C, 45 min
2. CuBr Me$_2$S, THF, –78°C, 30 min.

3 *p*-ClC$_6$H$_4$COCl, THF, –78°C to rt
18 h, 78%

46

47

catecholborane, tol., 99%

48

1. HBF$_4$·Et$_2$O, CH$_2$Cl$_2$, –60°C
2. 5 equiv. **49**, CH$_2$Cl$_2$,–60°C, 86%

49 —CO$_2$*n*-Bu

50

1. pyridine, reflux, 92%
2. 2 M HCl, 50°C, 4 h, 86%

—CO$_2$*n*-Bu

• 2HCl

—CO$_2$H

(*S*)-**4**

Scheme 7. The Corey asymmetric synthesis of (*S*)-cetirizine dihydrochloride, (*S*)-**4**.

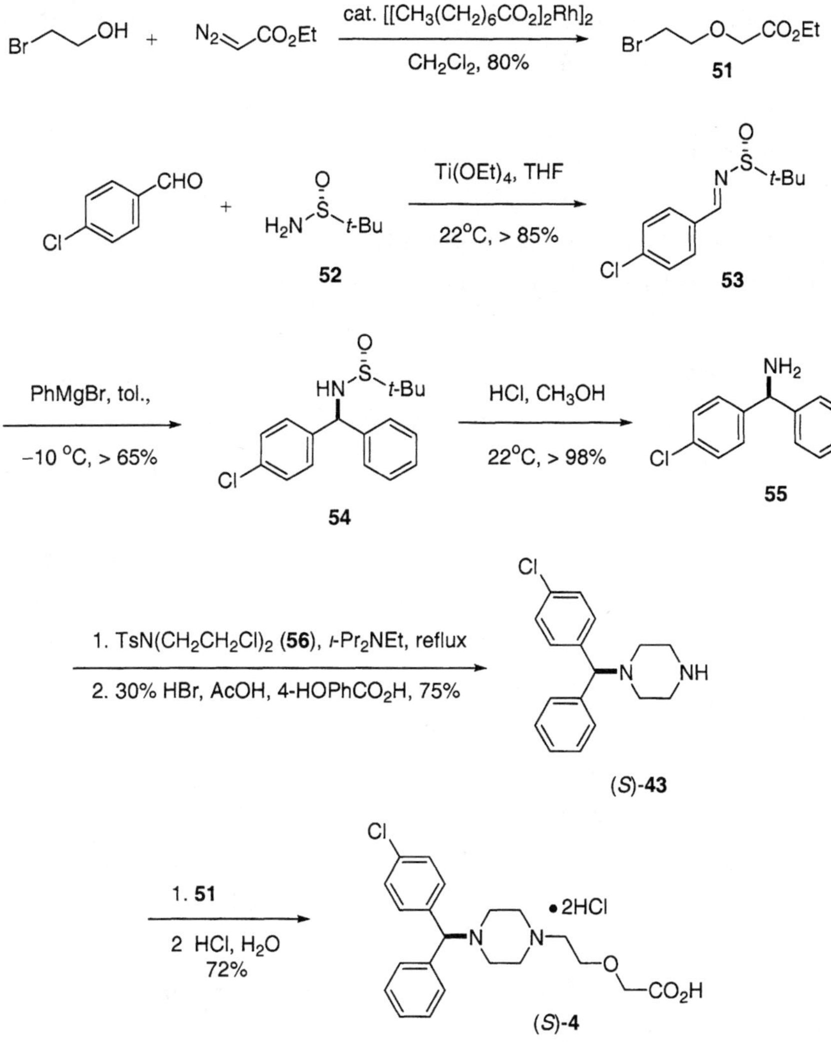

Scheme 8. The Sepracor synthesis of (*S*)-cetirizine dihydrochloride, (*S*)-**4**.

The first described synthesis of the enantiomeric cetirizine employed resolution of a (±)-chlorobenzhydrylamine as the salt with tartaric acid.[26] Later, an asymmetric synthesis was reported by the Corey group in 1996 (Scheme 7). The pivotal step involved a chiral oxazaborolidine (CBS)-catalyzed reduction of an unsymmetrical chlorobenzophenone with a π-chromium tricarbonyl group serving as an effective stereocontroller.[27]   Therefore, lithiation of benzene chromium tricarbonyl (**45**) was followed by treatment with CuBr·Me$_2$S to yield the corresponding organocopper reagent, which was then treated with 4-chlorobenzoyl chloride to give benzophenone **46** in 78% yield. Addition of a toluene solution of ketone **46** to 0.15 equivalent of oxazaborolidine **47** and 2 equivalents of catechlborane afforded alcohol **48** in 99% yield and 98% ee. In a Nicholas reaction-like transformation, optically pure chromium tricarbonyl-benzylic alcohol **48** underwent a S$_N$1 reaction with *retention* of configuration. The configurational stability was achieved because the carbocation intermediate was stabilized by Cr(CO)$_3$. Thus, in the presence of 2 equivalents of tetrafluoroboric acid-diethyl ether complex, **48** was treated with piperizine **49** to assemble adduct **50**.   De-metallation by refluxing piperizine **49** in pyridine was followed by hydrolysis to deliver (*S*)-cetirizine dihydrochloride (**4**).

Although this synthesis is evidently efficient, the use of the heavy atom, Cr, precludes it from industrial production purposes.

In the Sepracor synthesis of chiral cetirizine dihydrochloride (**4**), the linear side-chain as bromide **51** was assembled via rhodium octanoate-mediated ether formation from 2-bromoethanol and ethyl diazoacetate (Scheme 8).   Condensation of 4-chlorobenzaldehyde with chiral auxiliary (*R*)-*t*-butyl sulfinamide (**52**) in the presence of Lewis acid, tetraethoxytitanium led to (*R*)-sulfinimine **53**.   Addition of phenyl magnesium bromide to **53** gave rise to a 91 : 9 mixture of two diastereomers where the major diasteromer **54** was isolated in greater than 65% yield. Mild hydrolysis conditions were applied to remove the chiral auxiliary by exposing **54** to 2 N HCl in methanol to provide (*S*)-amine **55**. Bisalkylation of (*S*)-amine **55** with dichloride **56** was followed by subsequent hydrolysis to remove the tosyl amine protecting group to afford (*S*)-**43**. Alkylation of (*S*)-piperizine **43** with bromide **51** produced (*S*)-cetirizine ethyl ester, which was then hydrolyzed to deliver (*S*)-cetirizine dihydrochloride, (*S*)-**4**.

§5.5   References

1.   Azatadine, Villani, F. J.; Caldwell, W. US 3326924 (**1967**).

2. Loratadine, Villani, F. J.; Magatti, C. V.; Vashi, D. B.; Wong, J.; Popper, T. L. *Arzneimittel—Forsch* **1986**, *36*, 1311–1314.

3. Kling, J. *Modern Drug Discovery* **1999**, *2,* 49–54.

4. Loratadine, Villani, F. J. US 4282233 (**1981**).

5. Loratadine, Clissold, S. P.; Sorkin, E. M.; Goa, K. L. *Drugs* **1989**, *37,* 42–57.

6. Loratadine process synthesis, Schumacher, D. P.; Murphy, B. L.; Clark, J. E.; Tahbaz, P.; Mann, T. A. *J. Org. Chem.* **1989**, *54,* 2242–2244.

7. Desloratadine, Agrawal, D. K. *Expert Opinion on Investigational Drugs* **2001**, *10*, 547–560.

8. Desloratadine, Graul, A.; Leeson, P. A.; Castañer, J. *Drugs Fut.* **2000**, *25*, 339–346.

9. Desloratadine, Schumacher, D. P. EP 208855 (**1987**).

10. Desloratadine, Villani, F. J.; Wong, J. K. US 4659716 (**1987**).

11. Carr, A. A.; Kinsolving, C. R. US 387217 (**1975**).

12. Carr, A. A.; Dolfini, J. E.; Wright, G. J. US 4254129 (**1981**).

13. Baltes, E.; de Lannoy, J.; Rodriguez, L. US 4525358 (**1985**).

14. King, C.-H.; Kaminski, M. A. WO 93/21156 (**1993**).

15. Kawai, S. H.; Hambalek, R. J.; Just, G. *J. Org. Chem.* **1994**, *59*, 2620–2622.

16. Krauss, R.; Strom, R. M.; Scortichini, C. L.; Kruper, W. J.; Wolf, R. A.; Carr, A. A.; Rudisill, D. E.; Panzone, G.; Hay, D. A.; Wu, W. W. WO 95/00480 (**1995**).

17. Patel, S.; Waykole, L.; Repič, O.; Chen, K.-M. *Synth. Commun.* **1996**, *26*, 4699–4710.

18. Castañer, J. *Drugs Fut.* **1996**, *21*, 1017–1021.

19. Markham, A.; Wagstaff, A. J. *Drugs* **1998**, *55*, 269–274.

20. QT-issue, DuBuske, L. M. *Clinical Therapeutics* **1999**, *21*, 281–295.

21. Citrizine, Wellington, K.; Jarvis, B. *Drugs* **2001**, *61*, 2231–2240.

22. Schwender, C. F.; Beers, S. A.; Malloy, E. A.; Cinicola, J. *Bioorg. Med. Chem. Lett.* **1996**, *6*, 311–314.

23. Milani, C.; Carminati, G. M.; Sovera, A. GB 2076403 (**1981**).

24. Cossement, E.; Gobert, J.; Bodson, G. GB 2225320 (**1990**).

25. Baltes, E.; de Lannoy, J.; Rodriguez, L. US 4525358 (**1985**).

26. Opalka, C. J.; D'Ambra, T. E.; Faccone, J. J.; Bodson, G.; Cossement, E. *Synthesis* **1995**, *36*, 766–768.

27. Corey, E. J.; Helal, C. J. *Tetrahedron Lett.* **1996**, *37*, 4837–4840.

28. Pflum, D. A.; Wilkinson, H. S.; Tanoury, G. J.; Kessler, D. W.; Kraus, H. B.; Senanayake, C. H.; Wald, S. A. *Org. Process Res. Dev.* **2001**, *5*, 110–115.

29. Pflum, D. A.; Krishnamurthy, D.; Han, Z.; Wald, S. A.; Senanayake, C. H. *Tetrahedron Lett.* **2002**, *43*, 923–926.

Chapter 6. Cosmeceuticals:
Isotretinoin (Accutane®), Tazoratene (Tazorac®), Minoxidil (Rogaine®), and Finasteride (Propecia®)

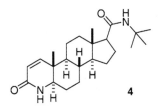

USAN Isotretinoin
Trade Name: Accutane®
Hoffman-La Roche
Launched 1982
M.W. 300.44

1

USAN: Tazarotene
Trade Name: Tazorac®
Allergan
Launched 1996
M.W. 351.47

2

USAN Minoxidil
Trade Name: Rogaine®
Upjohn/Pharmacia
Launched 1986
M W. 209.25

3

USAN: Finasteride
Trade Name· Propecia®
Merck
Launched: 1997
M.W. 372.54

4

Cosmeceuticals are substances that are applied to skin or hair but do not modify its structure and function.[1] One aspect that differentiates cosmeceuticals from cosmetics is that most, if not all, of them originate from *bona fide* medicinal chemistry programs. In this chapter, four representative cosmeceuticals are discussed: isotretinoin (**1**) and tazarotene (**2**) for acne; and minoxidil (**3**) and finasteride (**4**) for hair growth, respectively. Ironically, in spite of their market successes, neither minoxidil nor finasteride were initially discovered for their marketed use. Minoxidil was first synthesized as a potassium channel opener (KCO) for the treatment of hypertension, while finasteride was originally prepared as a testosterone-5α-reductase inhibitor for the treatment of benign prostatic hyperplasia (BPH).

§6.1   Isotretinoin (Accutane®)

§6.1.1 Introduction[2–6]

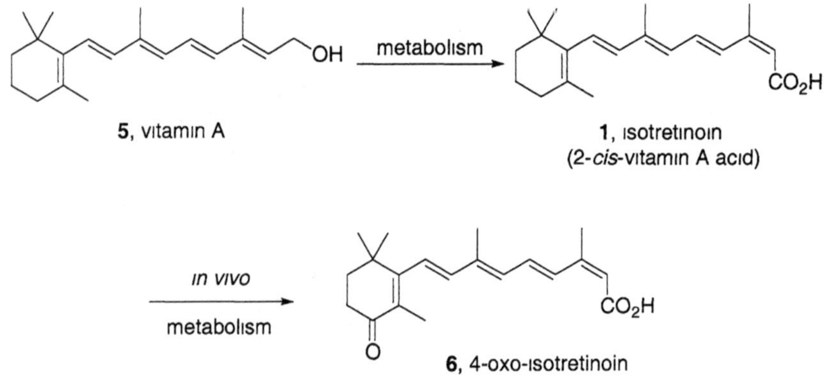

Fig. 1. Isotretinoin (**1**), a metabolite of vitamin A (**5**).

Isotretinoin (**1**, Accutane®), which is also known as 2-*cis*-vitamin A acid or 13-*cis*-retinoic acid is a naturally occurring metabolite of vitamin A (**5**, Fig. 1). It is an orally active drug for the treatment of severe refractory acne that functions by reducing sebaceous gland size and sebum production. *In vivo* studies in hamsters show that isotretinoin (**1**) causes atrophy of the sebaceous component of the flank organ by a mechanism not involving androgen inhibition. The major metabolite in the systemic circulation in man following oral administration is 4-oxo-isotretinoin (**6**).

Isotretinoin (**1**) works extremely well for severe acne, and is efficacious for 75% of all patients. However, like all oral retenoids, isotretinoin (**1**) has a spectrum of toxicity. Most prevalent of all is its profound teratogenic properties. The risk of a major congenital abnormality in the first trimester in pregnant women is increased by 25-fold. Therefore, it is mandatory for women of child-bearing potential who are taking isotretinoin (1) to avoid pregnancy. Another reported side effect of isotretinoin (1) is depression. Thus, patients should be carefully monitored with regard to mood change during administration of this drug.

§6.1.2 Synthesis of isotretinoin[7–8]

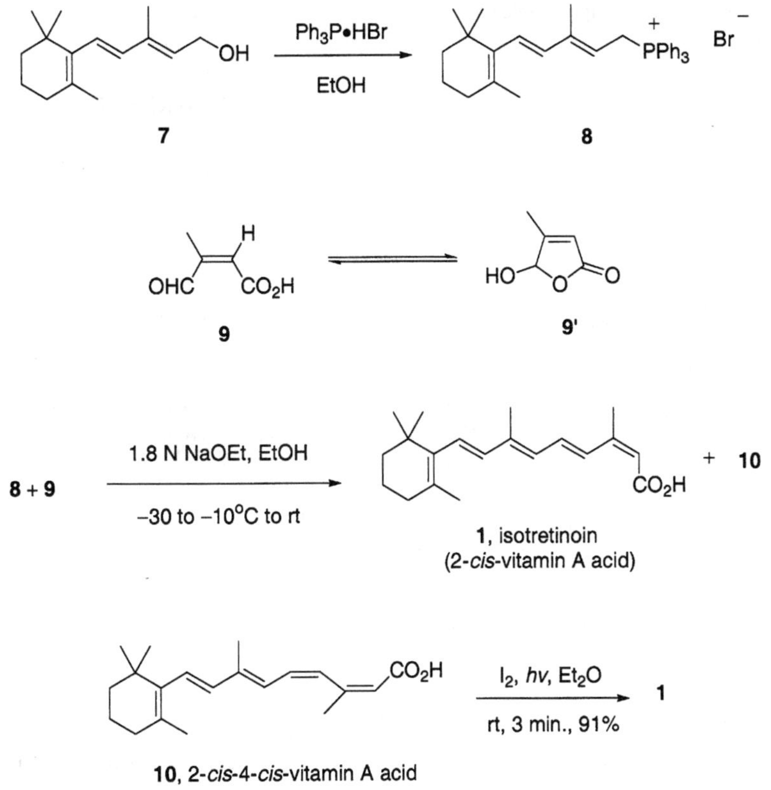

Scheme 1. Synthesis of isotretinoin (1).

As shown in Scheme 1, isotretinoin (1) was first synthesized by Garbers *et al.* in 1968 utilizing a key Wittig condensation.[7] Phosphonium salt **8** was prepared from direct treatment of vinyl β-ionol **7** with triphenylphosphonium bromide (Ph₃P•HBr) in ethanol. Subsequent addition of an excess of sodium ethoxide to **8** was followed by an ethanol solution of *cis*-β-formyl crotonic acid (**9**) to produce isotretinoin (**1**, 2-*cis*-vitamin A acid, or 13-*cis*-retinoic acid) and 2-*cis*-4-*cis*-vitamin A acid (**10**) in a 1:6 ratio. During the reaction, **9** was isomerized from hydroxybutenolide **9'** within 30 min. by treatment with ethanolic sodium ethoxide. Although the Wittig condensation gave the undesired 2-*cis*-4-*cis*-vitamin A acid (**10**) as the major product, it was conveniently converted to isotretinoin (**1**) *via* photoisomerization. Thus, a solution of **10** and iodine in ether was set aside at room temperature for 3 min in diffuse light to provide the isomerized isotretinoin (**1**) in 91% yield as orange red needles.

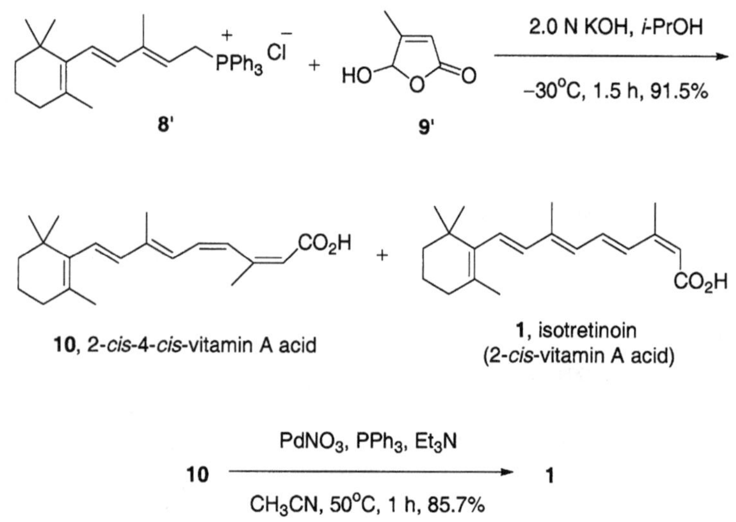

Scheme 2. The Hoffmann-La Roche synthesis of isotretinoin (**1**).

In a 1985 patent by Hoffmann-La Roche, the Wittig condensation was also the crucial step in assembling isotretinoin (**1**, Scheme 2).[8] Under the optimized conditions, 1.03 equivalents of phosphonium salt **8'** was condensed with 1 equivalent of hydroxybutenolide **9'** in the presence of 1.25 equivalents of 2 N KOH in isopropanol at –30°C for 1 to 1.5 h. The product (91.5% total yield) consisted of 75.9% of 2-*cis*-4-*cis*-vitamin A acid (**10**) and 16.7% of isotretinoin (**1**). Without separation, the mixture of **10** and **1** was subjected to palladium-catalyzed isomerization conditions: the mixture was heated at 50°C for 1 h in acetonitrile in the presence of 0.10 mol% of palladium(II) nitrate, four equivalents (based on palladium nitrate) of triphenylphosphine and 2 equivalents (also based on palladium nitrate) of triethylamine. As a consequence, isotretinoin (**1**) was obtained in 85.7% yield.

§6.2   Tazarotene (Tazorac®)
§6.2.1 Introduction[9–15]

Systemically administered retinoids such as isotretinoin (**1**, Accutane®) have several disadvantages such as a relatively narrow therapeutic index and a variety of toxic effects including teratogenicity. Topically administered retinoids may avoid some of those drawbacks. For instance, tazarotene (**2**, Tazorac®) is a topical receptor-selective retenoid that normalizes differentiation and proliferation of keratinocytes. Its major metabolite, tazarotenic acid (**11**), binds to retinoic acid receptors (RARs) with high affinity.

Tazarotene (**2**) is a prodrug. *A prodrug is defined as a pharmacologically inactive chemical derivative of a drug molecule that requires a transformation within the body in order to release the active drug.* In this particular case, the ethyl ester moiety is readily hydrolyzed by an esterase on skin to release the active metabolite, tazarotenic acid (**11**). Tazarotene (**2**) has a half-life of 2–18 min. and is rapidly cleared on skin.

The pharmacokinetics and metabolism of tazarotene (**2**) is especially worth noting. Topical gel application provides the direct delivery of tazarotene into the skin. Ten hours after a topical application of 0.1% tazarotene gel to the skin, approximately 4–6% of the dose resides in the stratum corneum and 2% of the dose is distributed to the viable epidermis and dermis.[15] As depicted in Scheme 3, both tazarotene (**2**) and tazarotenic acid (**11**) undergo further metabolism to their corresponding sulfoxides **12** and **13**, respectively. Sulfoxides **12** and **13**, in turn, are even further oxidized to sulfones **14** and **15**, respectively. These very polar metabolites do not accumulate in adipose tissue, but are rapidly eliminated *via* both urinary and fecal pathways with a terminal half-life of approximately 18 h. A lesson learned here is that installation of a sulfide moiety promotes clearance because it is oxidized to polar metabolites that are rapidly cleared. As the consequence, the systemic exposure is minimized. Percutaneous absorption of tazarotene (**2**) led to a plasma concentration below 1 μg/L. The systemic bioavailability of tazarotene (2) is thus low, approximately 1% after single and multiple topical applications to healthy skin.

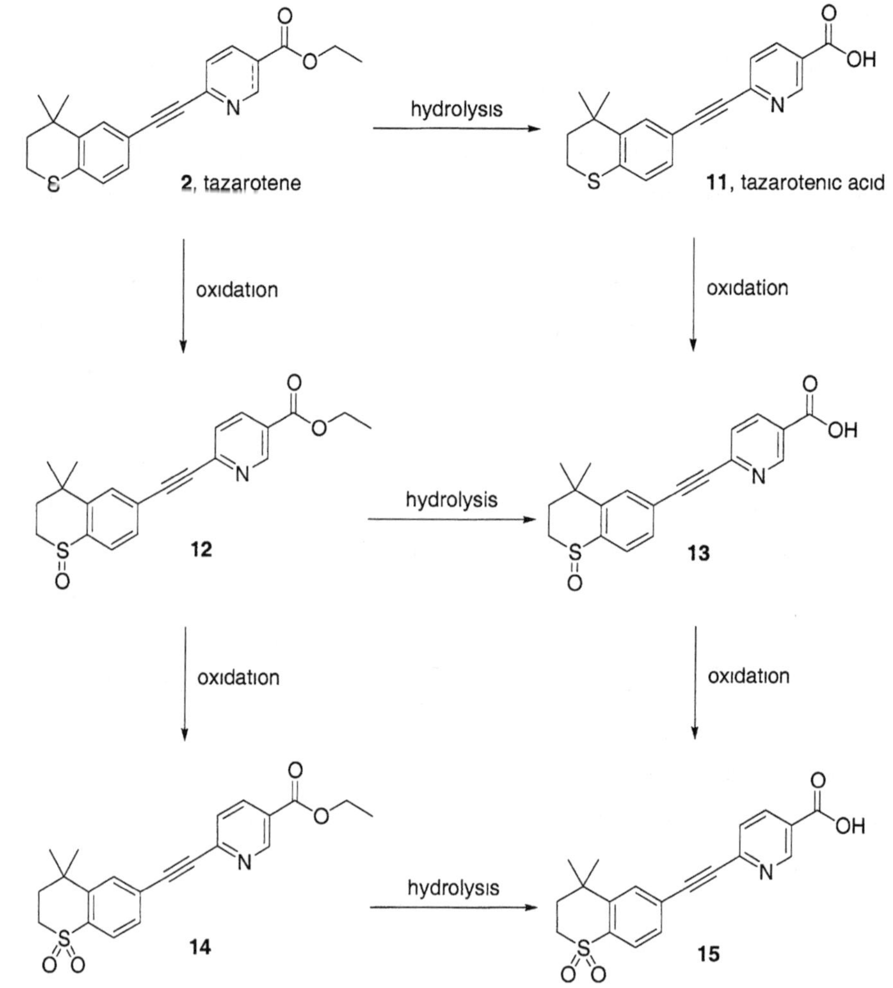

Scheme 3. Known metabolic pathways of tazarotene (2).[15]

Since mini-pig skin is similar to human skin particularly as it pertains to percutaneous absorption, the percutaneous absorption of tazarotene (2) has been carried out using mini-pigs.[14] In mini-pigs, high topical doses of tazarotene (2) produced only reversible local irritation, while lower doses were well tolerated topically.   More importantly, after one year of chronic dosing in mini-pigs there were no systemic toxic effects observed even at the highest possible dose which was limited by topical irritation.

§6.2.2 Synthesis of Tazarotene[16–18]

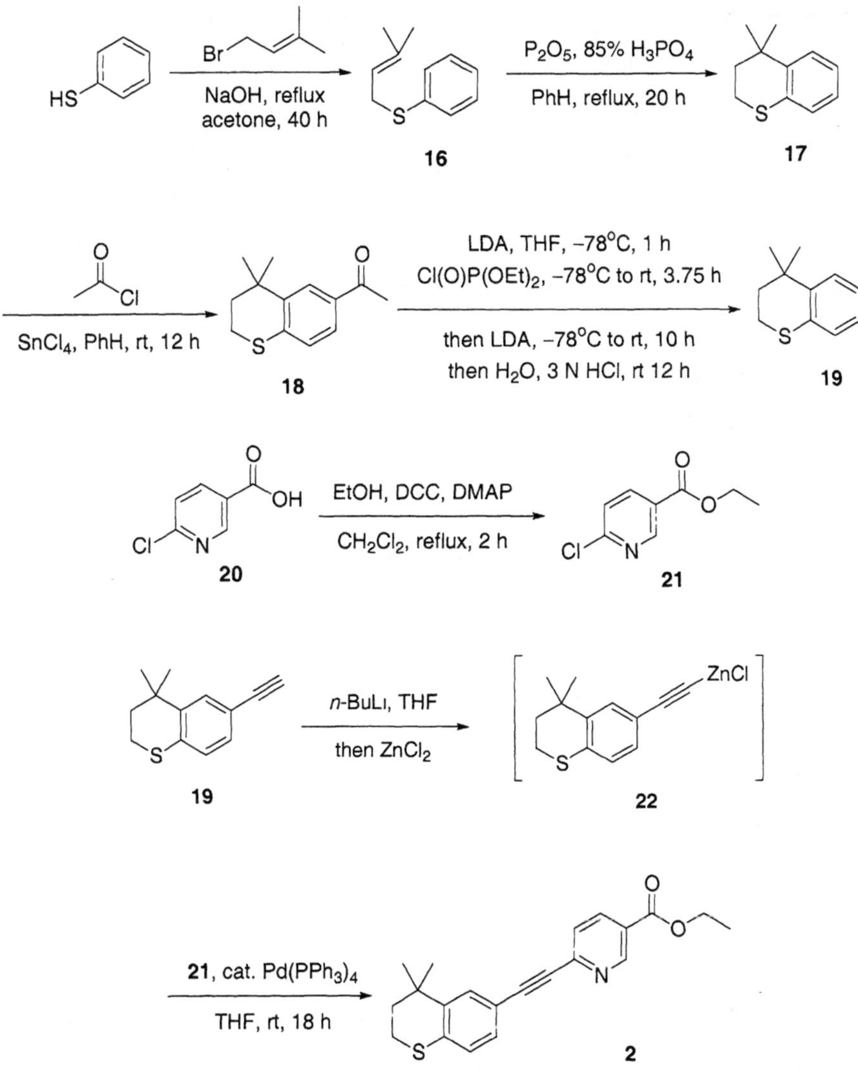

Scheme 4. The Allergan synthesis of tazarotene (2).

The key operation of the Allergan synthesis of tazarotene (**2**) is a Negishi coupling of alkynylzinc chloride **22** with α-chloropyridine **21** (Scheme 4).[16–18] Thus sulfide **16** was assembled *via* an S_N2 displacement of 1-bromo-3-methyl-but-2-ene with thiophenol with the aide of NaOH. Strong acid-promoted cyclization of **16** was achieved using P_2O_5/H_3PO_4 to afford 4,4-dimethylthiochroman (**17**). A Friedel–Crafts acylation of **17** was carried out using acetyl chloride in the presence of stannic chloride to produce methyl ketone **18**. Transformation of methyl ketone **18** to terminal alkyne **19** was then accomplished *via* the formation of an enol phosphate intermediate. On the other hand, DCC-mediated esterification of 6-chloronicotinic acid (**20**) led to formation of α-chloropyridine **21**. With two key intermediates, **21** and **19**, in hand, alkynylzinc chloride **22** was generated *in situ* by treatment of terminal alkyne **19** with *n*-BuLi followed by addition of $ZnCl_2$. Subsequently, the Negishi coupling of **22** with α-chloropyridine **21** using $Pd(Ph_3P)_4$ as the catalyst delivered tazarotene (**2**).

§6.3   Minoxidil (Rogaine®)

§6.3.1 Introduction[19–26]

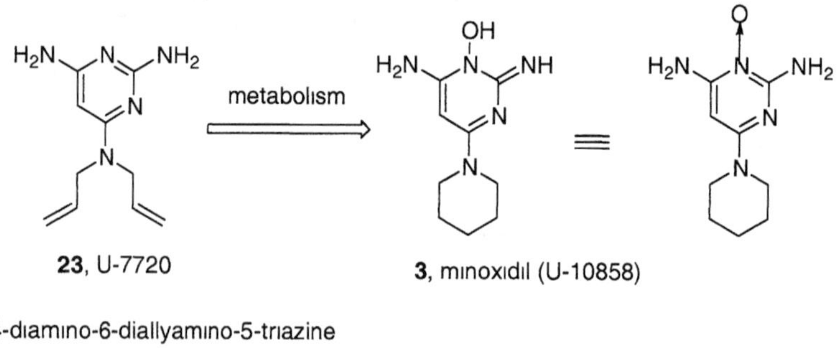

23, U-7720   **3**, minoxidil (U-10858)

2,4-diamino-6-diallyamino-5-triazine

Fig. 2. Evolution of U-7720 (**23**) to minoxidil (**3**).

N',N'-Diallyl-pyrimidine-2,4,6-triamine, U-7720 (**23**) was initially, developed as a potent antihypertensive agent. Later, its metabolite minoxidil (**3**) was found to be efficacious as both an antihypertensive orally and for hair growth topically. The systemic and local side effects of topical minoxidil (**3**) are essentially non-existent.

Minoxidil (3), 2,4-diamino-6-piperidinopyrimidine-3-oxide, is a potassium channel opener (KCO). It is a prodrug and is activated *via* sulfation *in vivo*. It was initially prepared as a potent vasodilator for the treatment of severe hypertension. When administered orally, a reversible hypertrichosis (excess hair growth) of the face, arms, and legs takes place in approximately 80% of the patients. However, side effects from the systemic administration of the drug include pigmentation and coarsening of facial features, fluid retention, tachycardia, nausea, fatigue, dyspnea, and gynecomastia. In contrast, there have been few reported side effects from the topical application of minoxidil (3). In a study, the percutaneous absorption and excretion of both a 1% and 5% solution of minoxidil (3) labeled with ^{14}C was investigated. While the urinary excretion of radioactivity was low, ranging from 1.6% to 3.9% of applied dose, no fecal radioactivity was detected.[24] The majority of the recovery of the radioactivity was on the skin surface, scalp and pillow case washes and ranged from 41% to 45% of applied dose. No adverse reactions or notable abnormalities were noted in the subjects during the studies. Although minoxidil (3) is poorly absorbed through the skin, systemic doses in the range of 2.4% to 5.4% mg/day can be expected if application is made to the entire scalp.

§6.3.2 Synthesis of minoxidil[27–29]

The initial patent describing the Upjohn synthesis of minoxidil (2) was issued in 1972 (Scheme 5).[27] The synthesis began with the assembly of phenoxypyrimidine 25 from the S$_N$Ar displacement of 4-chloropyrimidine-2,6-diamine (24) with 2,4-dichlorophenol. The aryl ether installed here serves as a leaving group in the last step, and is more tolerant of the ensuing oxidation step. Therefore, oxidation of phenoxypyrimidine 25 gave 6-amino-4-(2,4-dichlorophenoxy)-2-imino-2*H*-pyrimidin-1-ol (26), which was then condensed with piperidine in sealed tube at 250 °C to deliver minoxidil (3).

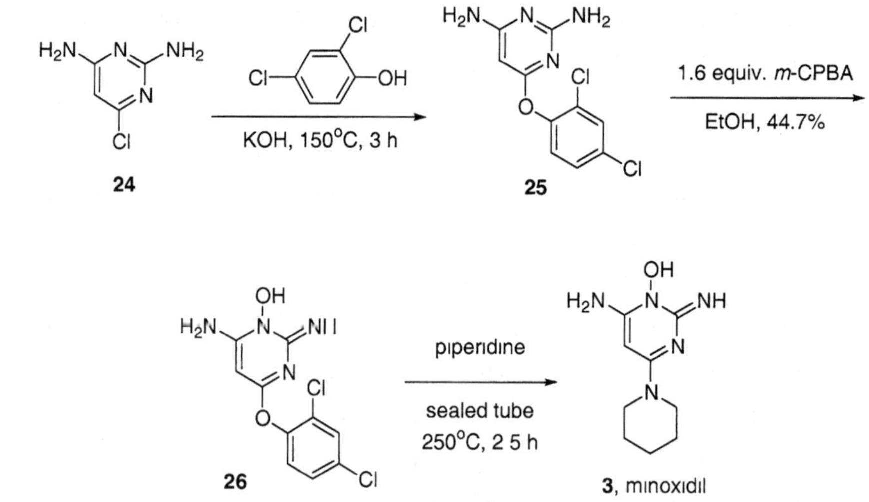

Scheme 5. The first Upjohn synthesis of minoxidil (2).[27]

The second patent by Upjohn describes an improved route, obviating the use of the sealed tube reactor (Scheme 6). Therefore, 6-amino-4-chloro-2-imino-2H-pyrimidin-1-ol (28) was prepared from the oxidation of 4-chloro-pyrimidine-2,6-diamine (27). The crude product was extracted with boiling acetonitrile to give pure 28 in 44.7% yield. Refluxing 28 with excess of piperidine for 1.5 h then afforded minoxidil (3) after extraction with boiling acetonitrile.

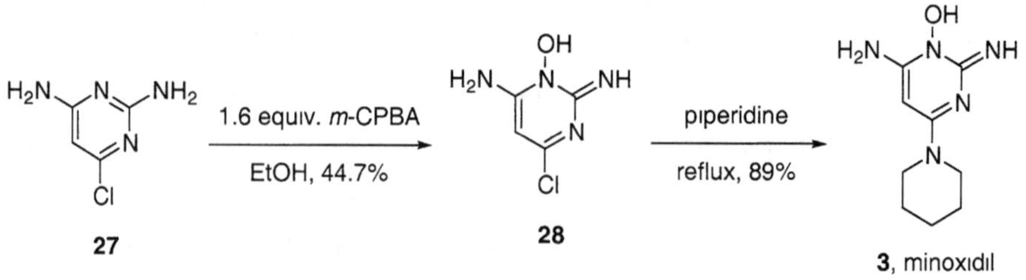

Scheme 6. The second Upjohn synthesis of minoxidil (3).[28]

These two syntheses have one theme in common — they all involve modification of a pre-existing pyrimidine N-oxide. However, the third Upjohn synthesis of minoxidil (3), on the other hand, entails the cyclization of a linear intermediate 33 as the key operation (Scheme 7).[29]   To this end, cyanoacetamide **30** was produced from the condensation of ethyl cyanoacetate (**29**) with piperidine.   *O*-Methylation of cyanoacetamide **30** was accomplished using either methyl fluorosulfonate or trimethyloxonium fluroborate to give enol ether **31**. Subsequent treatment of **31** with cyanamide in an alcoholic solvent then led to cyanoiminopropionitrile **32**. When **32** was treated with hydroxylamine hydrochloride, it is reasonable to assume that it attacks the more electron-deficient *N*-cyano group rather than the relatively electron-rich aliphatic nitrile group. As a consequence, minoxidil (**3**) was obtained in approximately 48% yield in 2 steps presumably *via* linear intermediate **33**.

In comparison to the chloro *N*-oxide routes, the third route is superior because it gives higher overall yield, requires less chromatography, and is substantially more economical.

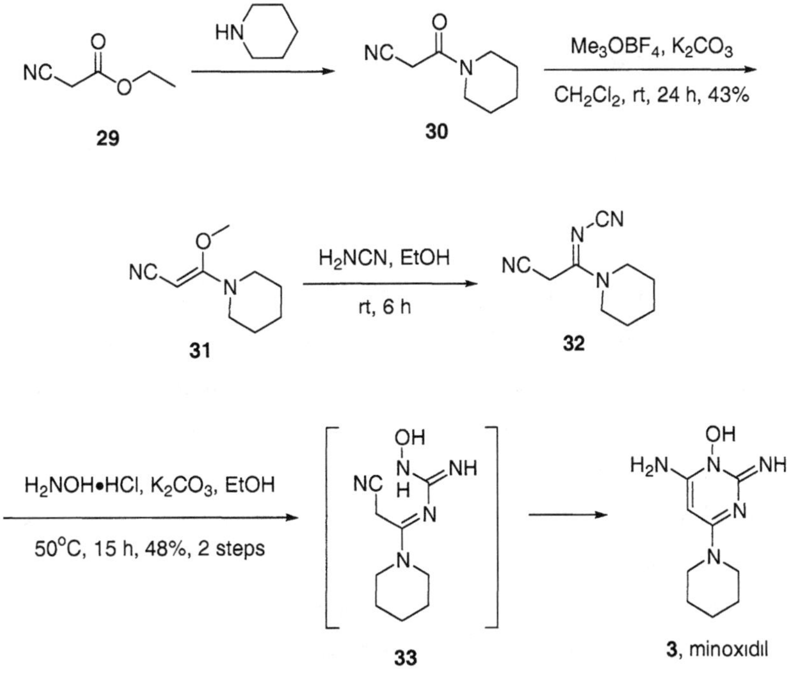

Scheme 7. The third Upjohn synthesis of minoxidil (**3**).[29]

§6.4   Finasteride (Propecia®)

§6.4.1 Introduction[30-38]

The correlation between male pattern baldness and the androgens has been long established. For instance, prepubertal gonadectomy of males prevents recessioin of scalp hair. However, the approach to stop and reverse male pattern baldness using systemic antiandrogens (androgen blockade, or androgen receptor antagonism) has not been pursued because of the feminizing effect and interference with male sexual functions. On the other hand, the functions of intracellular androgen receptor and 5α-reductase include regulation of hormone action in androgen-sensitive cells. 5α-Reductase converts testosterone (T, **34**) to a more potent androgen, 5α-dihydrotestosterone (DHT, **35**).

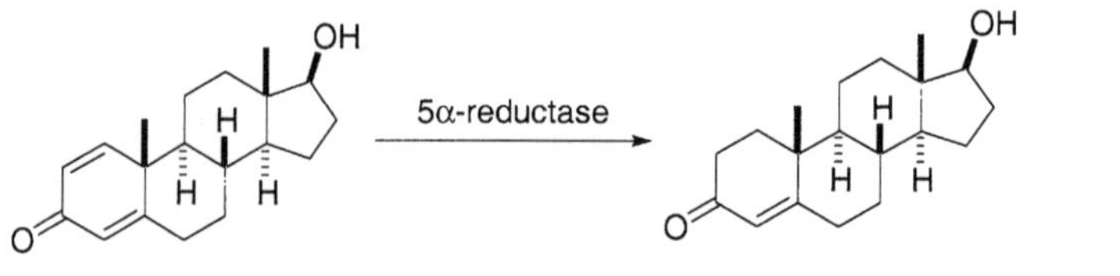

34, testosterone   **35**, 5α-dihydrotestosterone

Fig.3. Testosterone (**34**) and 5α-dihydrotestosterone (**35**).

Finasteride (**4**), an azasteroid, was originally developed as a specific competitive inhibitor of 5α-reductase for the therapeutic indication of benign prostatic hyperplasia (BPH) with trade name as Proscar®. It has no binding affinity for androgen receptor sites and finasteride (**4**) itself possesses no androgenic, anti-androgenic, or other steroid hormone-related properties. Therefore, one significant property of finasteride (**4**) is lowering levels of DHT without interfering with testosterone (**34**) levels. Merck used balding male and female monkeys, stumptail macaques (*Macaca arctoides*), to test the effects of finasteride (**4**, Propecia®) on hair growth, hair cycle stage, and serum testosterone and dihydrotestosterone.[32] Before that, Upjohn also used balding stumptail macaques to study the effects of oral finasteride (**4**) alone or in combination with topical minoxidil (**3**).[33] It was found that finasteride (**4**) increased the hair weight for four of five monkeys, and it was more efficacious in combination with minoxidil (**3**). In clinical trials, thanks to the excellent bioavailability (average 63% in humans) of finasteride (**4**), it is efficacious at a dose of 5 mg/day for androgenic alopecia.[34,35] It was found that finasteride (**4**) lowers levels of DHT without interfering with T levels on scalp skin.

In a study of the metabolism of finasteride (**4**) using both healthy volunteers and rat heptic microsomes, five major metabolites were identified as ω-hydroxy finasteride (**36**), 6α,ω-dihydroxy finasteride (**37**), finasteride-ω-oic acid (**38**), its corresponding methyl ester **39**, and 6α-OH finasteride (**40**), respectively.[36–38]

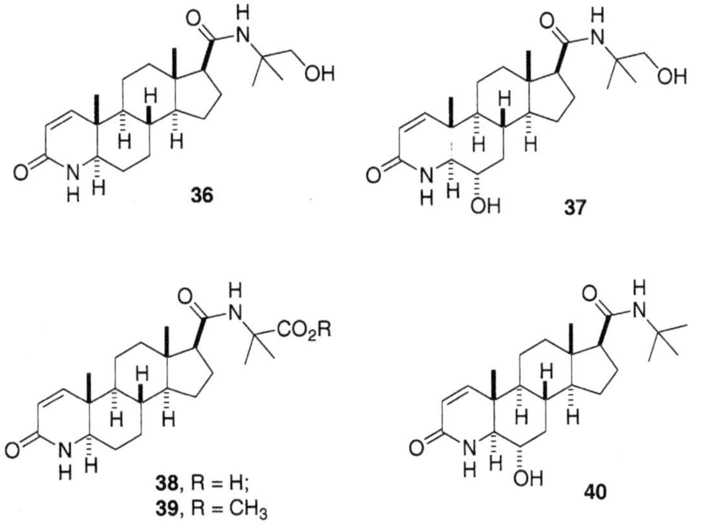

Fig.4. Five major metabolites of finasteride (**4**).

§6.4.2 Synthesis of finasteride[39–44]

In a Merck patent published in 1988,[41] the synthesis of finasteride (**4**) utilized 3-oxo-4-etien-20-oic acid **41** as a starting point. Periodate olefin cleavage of **41** was achieved using NaIO₄ and KMnO₄ to give dioic acid **42**. Ring closure with ammonia at high temperature afforded ene lactam **43**, which was selectively hydrogenated using a platinum catalyst to afford 5α-4-azasteroid **44**. Amide **45** was then assembled from a coupling of acid **44** and *t*-butyl amine using a combination of DCC and 1-

hydroxybenztriazole.   Finally, dehydrogenation of **45** was carried out using benzeneseleninic anhydride in refluxing chlorobenzene to fashion finasteride (**4**).

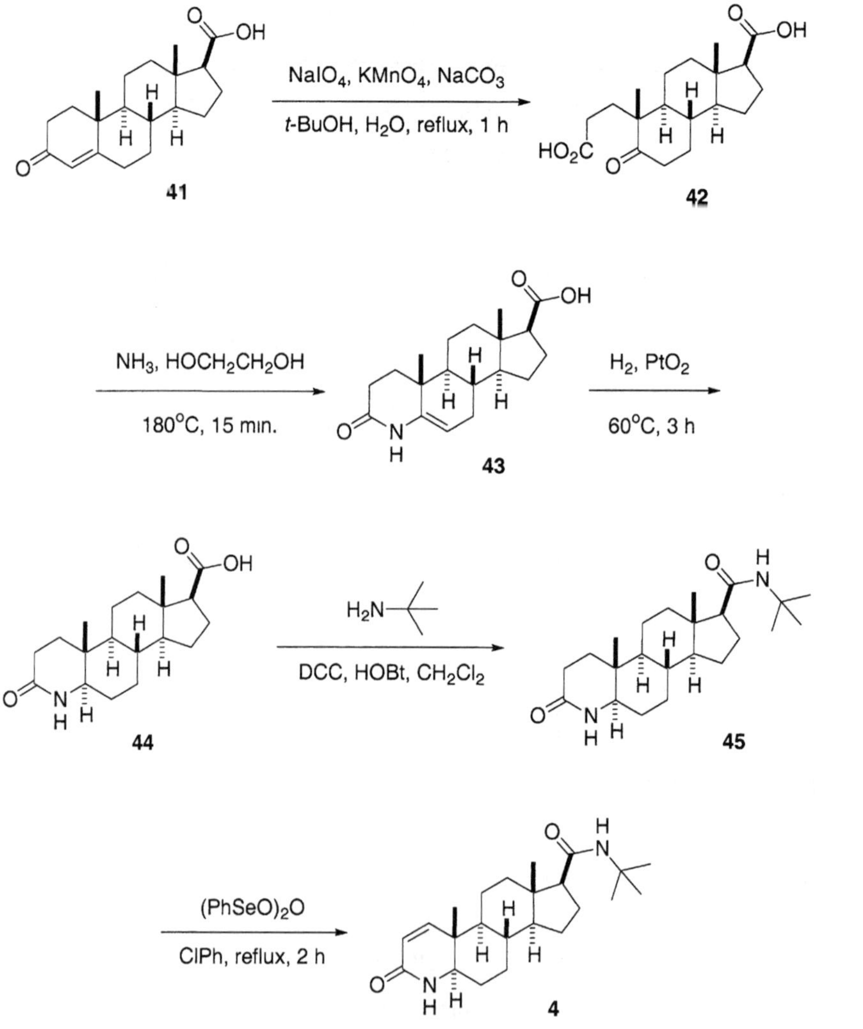

Scheme 5. The Merck synthesis of finasteride (**4**).

The interesting transformation **45** → **4** was also accomplished by refluxing **45** with DDQ (2,3-dichloro-5,6-dicyano-1,4-benzoquinone) and BSTFA [bis(trimethylsilyl)-trifluoroacetamide] in dioxane.[42]   NMR spectroscopy studies suggested the following mechanistic pathway: *O*-Silyl imidate **46** was produced under the reaction conditions at room temperature. Nucleophilic addition of **46** to DDQ led to adduct **47**, which was subsequently silylated to another *O*-silyl imidate **48**. At elevated temperature, *O*-silyl imidate **48** expelled 4,5-dichloro-3,6-bis-trimethylsilanyloxy-phthalonitrile (**49**) to give dehydrogenated lactam, finasteride (**4**). The conversion seems to be general for similar Δ^1-azasteroids.

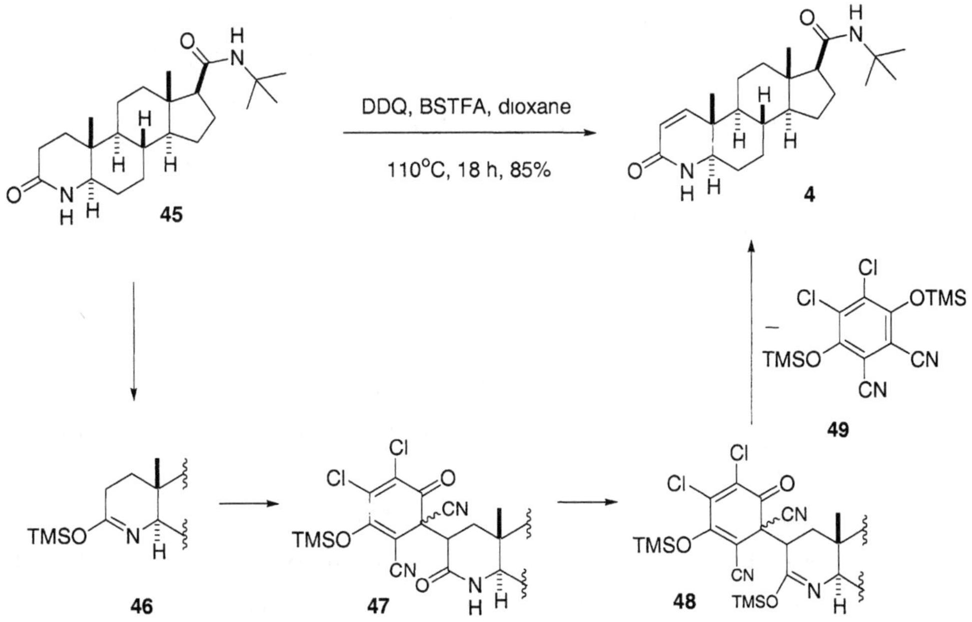

Scheme 6. DDQ-BSTFA-mediated dehydrogenation of azasteroid **5**.

Furthermore, the transformation **44** → **45** was carried out using an acylimidazole intermediate to give an excellent yield of the observed product.[43] Thus, acylimidazole **50** was prepared, via the action of carbonylimidazole in THF, as a stable crystalline compound. Subsequently, *t*-butylaminomagnesium bromide was produced by treatment of *t*-butylamine with ethylmagnesium bromide. Finally, refluxing the mixture of acylimidazole **50** and *t*-butylaminomagnesium bromide in THF for 18 h gave the amide **45**.

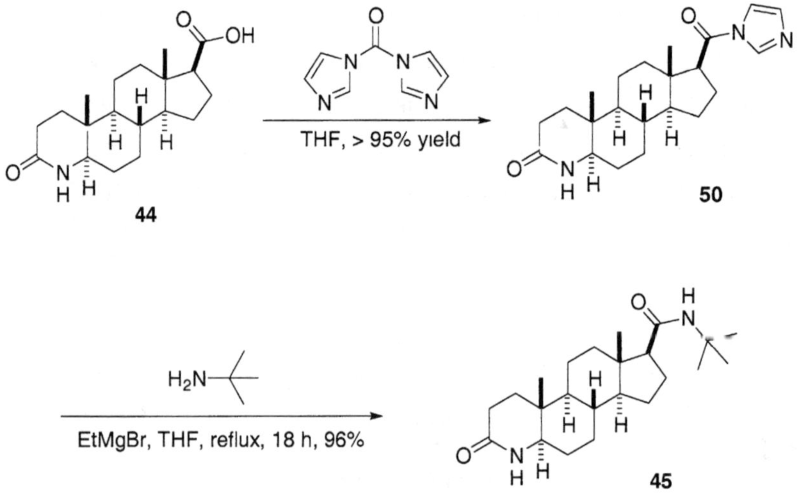

Scheme 7. Amide synthesis *via* acylimidazole intermediate.

In order to quantitatively determine the level of finasteride at the picogram level in human plasma using isotope-dilution gas chromatography mass spectrometry, Guarna *et al.* synthesized 5,6,6-[^{2}H$_3$]finasteride (**55**).[44] Known intermediate **44** was exchanged with deuterated water and then deuterated acetic acid to give intermediate **51**, which was hydrogenated *in situ* with 3 atm ^{2}H$_2$ to produce D$_3$-aza-steroid **52**. DDQ-BSTFA-mediated dehydrogenation of aza-steroid **52** afforded dehydrogenated lactam **53**. Activation of the carboxylic acid was realized by the subsequent treatment of **53** with 2,2'-dithiopyridine and triphenylphospine in toluene to give the thioester **54**. Reaction between **54** and *t*-butylamine then led to 5,6,6-[^{2}H$_3$]finasteride (**55**).

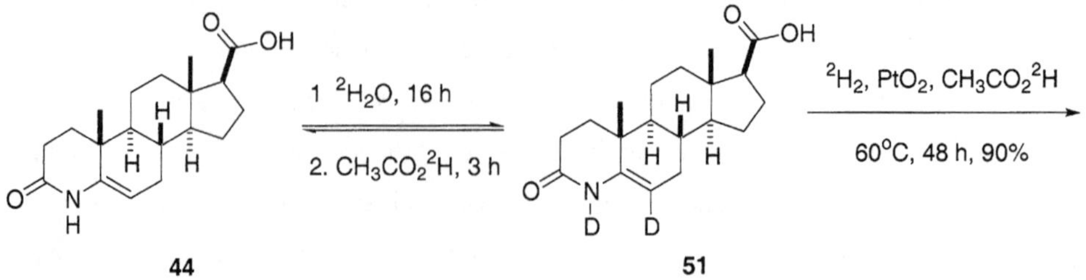

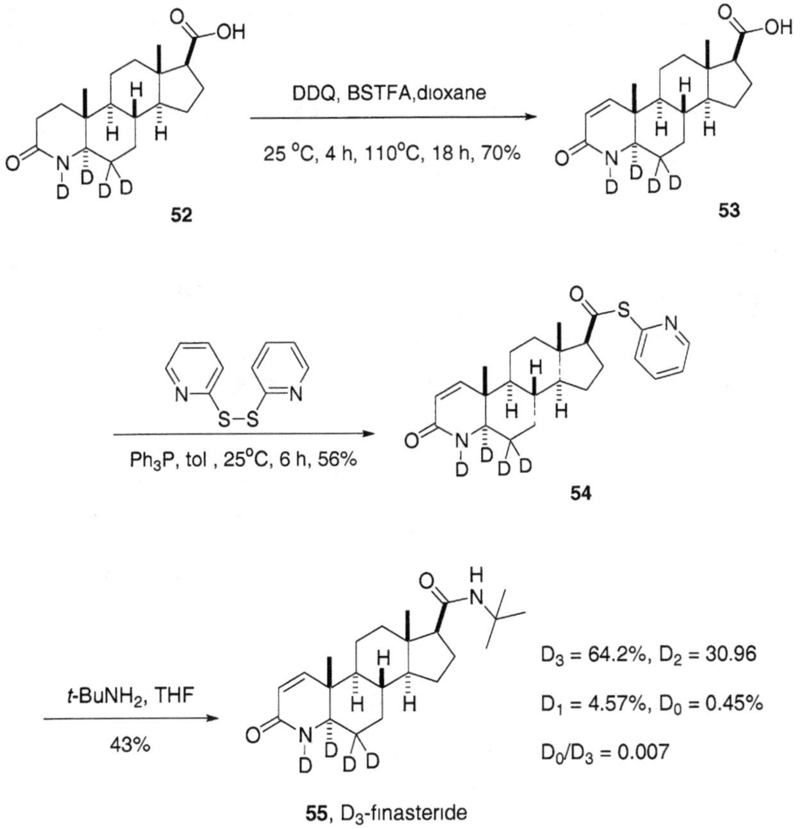

DDQ, BSTFA, dioxane

25 °C, 4 h, 110°C, 18 h, 70%

52   **53**

Ph₃P, tol , 25°C, 6 h, 56%

54

t-BuNH₂, THF

43%

$D_3 = 64.2\%$, $D_2 = 30.96$

$D_1 = 4.57\%$, $D_0 = 0.45\%$

$D_0/D_3 = 0.007$

55, D₃-finasteride

Scheme 8. Synthesis of D₃-finasteride **55**.

§6.5   References

1.   *Cosmeceuticals, Drugs vs. Cosmetics;* Elsner, P.; Maibach, H. I., Eds.; Marcel Dekker: New York, 2000.

2.   Ward, A.; Brogden, R. N.; Heel, R. C.; Speight, T. M.; Avery, G. S. *Drugs* **1984**, *28*, 6–37.

3.   Paust, J. *Pure Appl. Chem.* **1991**, *63*, 45–58.

4.   Orfanos, C. E.; Zouboulis, C. C.; Almond-Roesler, B.; Geilen, C. C. *Drugs* **1997**, *53*, 358–388.

5. Nagpal, S.; Chandraratna, R. A. S. *Current Pharmaceutical Design* **2000**, *6*, 919–931.

6. Niles, R. M. *Exp. Opinion Pharmacotherapy* **2002**, *3*, 299–303.

7. Garbers, C. F.; Schneider, D. F.; van der Merwe, J. P. *J. Chem. Soc. (C)* **1968**, 1982–1983.

8. Lucci, R. (Hoffmann-La Roche), US 4556518 (**1985**).

9. Ngo, J.; Leeson, P. A.; Castañer, J. *Drugs Fut.* **1997**, *22*, 249–255.

10. Menter, A. *J. Am. Acad. Derm.* **2000**, *43*, S31–5.

11. Foster, R. H.; Brogden, R. N.; Benfield, P. *Drugs* **1998**, *55*, 705–712.

12. Marks, R. *J. Am. Acad. Derm.* **1988**, *39*, S134–8.

13. Nagpal, S.; Chandraratna, R. A. S. *Current Pharm. Design* **1998**, *14*, S134–8.

14. Chandraratna, R. A. S. *Brit. J. Derm.* **1996**, *135*, 18–25.

15. Tang-Liu, D. D.; Matsumoto, R. M.; Usansky, J. I. *Clin. Pharmacokinetics* **1999**, *37*, 273–87.

16. Chandraratna, R. A. S. EP 0284288 (**1988**).

17. Chandraratna, R. A. S. US 5089509 (**1992**).

18. Chandraratna, R. A. S. WO 96/11686 (**1996**).

19. Sungurbey, K. *Drugs Fut.* **1977**, *11*, 383–386.

20. Campese, V. M. *Drugs* **1981**, *22*, 257–278.

21. Fenton, D. A. and Wilkinson, J. D. *Brit. Med. J.* **1983**, *287*, 1015–1017.

22. Weiss, V. C.; West, P.; Fu, T. S.; Robinson, L. A.; Cook, B.; Cohen, R. L.; Chambers, D. A. *Arch. Dermatol.* **1984**, *121*, 457–463.

23. Weiss, V. C.; West, D. P. *Arch. Dermatol.* **1985**, *121*, 191–192.

24. Franz, T. J. *Arch. Dermatol.* **1985**, *121*, 203–206.

25. De Viliez, R. L. *Arch. Dermatol.* **1985**, *121*, 382–383.

26. Anonymous, *Drugs Fut.* **1986**, *22*, 249–255.

27. Anthony, W. C.; Ursprung, J. J. US 3382247 (**1968**).

28. Anthony, W. C. US 3644364 (**1972**).

29. McCall, J. M.; TenBrink, R. E.; Ursprung, J. J. *J. Org. Chem.* **1975**, *40*, 3304–3306.

30. Prous, J.; Castañer, J. *Drugs Fut.* **1991**, *16*, 996–1000.

31. Sudduth, S. L.; Koronkowski, M. J. *Pharmacotherapy* **1993**, *13*, 309–329.

32. Rhodes, L.; Harper, J.; Uno, H.; Gaito, G.; Audette Arruda, J.; Kurata, S.; Berman, C.; Primka, R.; Pikounis, B. *J. Clin. Endocrinol. Metab.* **1994**, *79*, 991–996.

33. Diani, A. R.; Mulholland, M. J.; Shull, K. L.; Kubicek, M. F.; Johnson, G. A. *J. Clin. Endocrinol. Metab.* **1992**, *74*, 345–350.

34. Dallob, A. L.; Sadick, N. S.; Unger, W.; Lipert, S.; Geissler, L. A.; Greguire, S. L.; Nguyen, H. H.; Moore, E. C.; Tanaka, W. K. *J. Clin. Endocrinol. Metab.* **1994**, *79*, 703–706.

35. Moghetti, P.; Castello, R.; Magnani, C. M.; Tosi, F.; Negri, C.; Armanini, D.; Bellotti, G.; Muggeo, M. *J. Clin. Endocrinol. Metab.* **1994**, *79*, 1115–1121.

36. Ishii, Y.; Mukoyama, H.; Ohtawa, M. *Drug Metab. Disposition* **1994**, *22*, 79–84.

37. Carlin, J. R.; Christofalo, P.; Arison, B. H.; Berman, C.; Brooks, J. R.; Rasmusson, G. H.; Rosegay, A.; Stoner, E.; VandenHeuvel, J. A. *Pharmacologist* **1987**, *29*. (abstr.) 149.

38. Carlin, J. R.; Hoglund, P.; Eriksson, L. O.; Christofalo, P.; Gregoire, S. L.; Taylor, A. M.; Andersson, K. E. *Drug Metab. Disposition* **1992**, *20*, 148–144.

39. Rasmusson, G. H.; Reynolds, G. F.; Utne, T.; Jobson, R. B.; Primka, R. L.; Berman, C.; Brooks, J. R. *J. Med. Chem.* **1984**, *27*, 1690–1701.

40. Rasmusson, G. H.; Reynolds, G. F.; Steinberg, N. G.; Walton, N. E.; Patel, G. F.; Liang, T.; Cascieri, M. A.; Cheung, A. H.; Brooks, J. R.; Berman, C. *J. Med. Chem.* **1988**, *29*, 2298–2315.

41. Rasmusson, G. H.; Reynolds, G. F. US 4760071 (**1988**). *Same as* Rasmusson, G. H.; Reynolds, G. F. EP 285383 (**1988**).

42. Bhattachari, A.; DiMichele, L. M.; Dolling, U.-H.; Douglas, A. W.; Grabowski, E. J. J. *J. Am. Chem. Soc.* **1988**, *110*, 3318–3319.

43. Bhattachari, A.; Williams, J. M.; Dolling, U.-H.; Grabowski, E. J. J. *Synth. Commun.* **1990**, *30*, 2683–3319.

44.   Guarna, A.; Danza, G.; Bartolucci, G.; Marrucci, Dini, S.; Serio, M. *J. Chromatgr. — B, Biomed. Appl.* **1995**, *674*, 197–204.

Chapter 7. Antibacterials:
Ciprofloxacin (Cipro®), and Linezolid (Zyvox®)

USAN: Ciprofloxacin
Trade Name: Cipro®
Bayer
Launched: 1983
M.W 331 34

1

USAN: Linezolid
Trade Name: Zyvox®
Pharmacia
Launched: 2000
M.W. 337.35

2

Alexander Fleming discovered penicillin (**3**) in 1928 in England. However, it was not made commercially until 1941 when Florey and Chain devised the process that helped the American pharmaceutical industry to manufacture penicillin *via* deep-tank fermentation. Florey, Chain and Fleming shared the Nobel Prize in medicine in 1945. In 1932, Fritz Mietzsch and Josef Klarer at I. G. Farben in Germany synthesized Prontosil®, 2'4'diaminoazobenzene-4-sulfonamide (**4**). Gerhard Domagk, also at I. G. Farben, discovered **4**, the "sulfa drug" as an antibacterial against pneumonia and meningitis. He is remembered for having won the Nobel Prize for medicine in 1939 in addition to having cured his first patient, his daughter, using Prontosil®. That event heralded the beginning of the modern antibacterial era.

The arsenal of antibacterials available for the treatment of infectious diseases has expanded exponentially since then. Currently, as illustrated in Fig. 1, the repertoire includes the β-lactams (e.g. **3**); sulfa drugs (e.g. **4**); macrolides (e.g. **5**, erythromycin A), nitrofuran drugs (e.g. **6**, furazolidone), and many aminoglycosides.

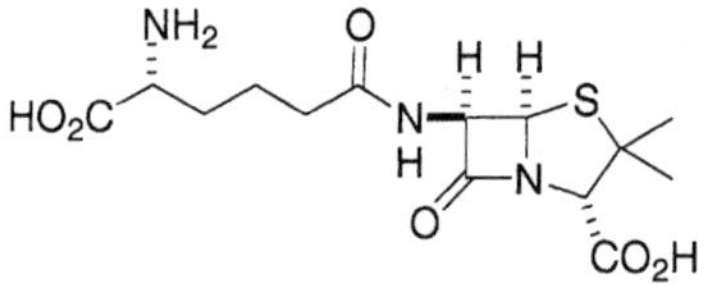

3, penicillin 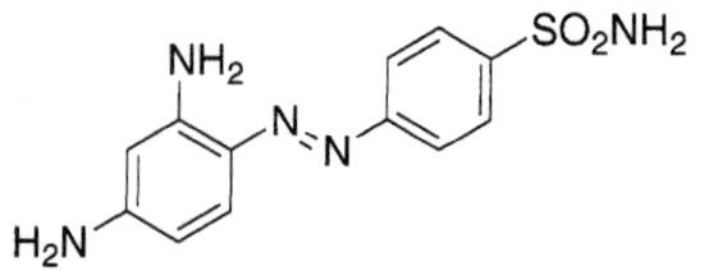   **4**, Prontosil

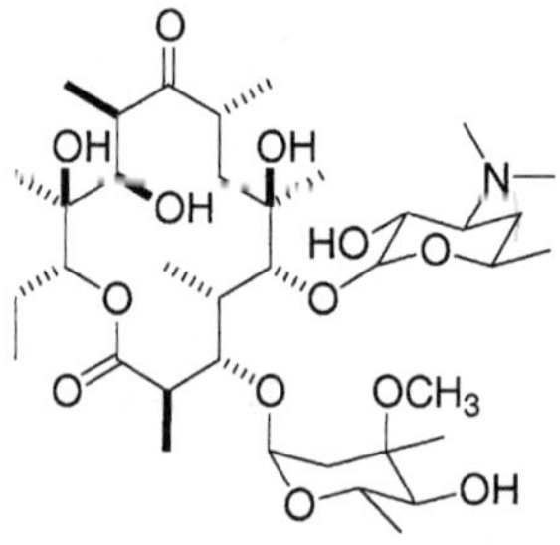

6, Furazolidone

5, Erythromycin A

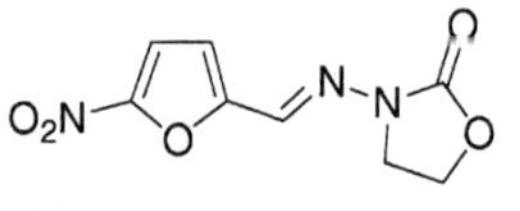

Fig. 1. Representative antibacterials.

§7.1 Ciprofloxacin (Cipro®)

§7.1.1 Introduction to ciprofloxacin (1)[1–7]

A very important category of antibacterial agents is the quinolones. In 1962, nalidixic acid (**7**) was first introduced by Lesher into clinical practice for urinary tract infections because it was excreted *via* urine in high concentration. Shortly after that, the quinolone antibacterials flourished, rendering thousands of 4-quinolones as represented by cinoxacin (**8**) and pipemidic acid (**9**). Quinolones **7–9** are considered as the first-generation quinolone antibacterials for their moderate activities against susceptible bacteria and poor pharmacokinetic (PK) properties. They possess oral activity against Gram-negative bacteria but suffer as a class in their inability to affect Gram-positive strains. Furthermore, the bioavailabilties are too low to treat systemic infections such as pneumonia and skin infections.

In the early 1980s, fluorinated quinolone antibacterials were discovered to possess better activity and PK properties (longer half lives and better oral efficacy). These so-called second-generation quinolone antibacterials are exemplified by norfloxacin (**10**), perfloxacin (**11**) and ciprofloxacin (**1**). They display a broader spectrum of antibacterial activity, increased potency, decreased potential for resistance and less toxicity. They have become the first line of attack for the clinical treatment of a variety of infectious diseases in contemporary medicine. In this chapter, ciprofloxacin (**1**) is chosen as an example to highlight the quinolone antibacterials partially because of its stardom in the wake of the bio-terrorism threat. It is worth mentioning that the third-generation of quinolone antibacterials are still being actively investigated because of the rapid development of resistance by bacteria towards existing antibacterial drugs. Examples of the third generation of quinolone antibacterials include fleroxacin (**12**), sparfloxacin (**13**) and tosufloxacin (**14**). They are endowed with sufficiently long half-lives to enable a once daily regimen, along with enhanced activities against a variety of bacteria.

First-generation quinolones

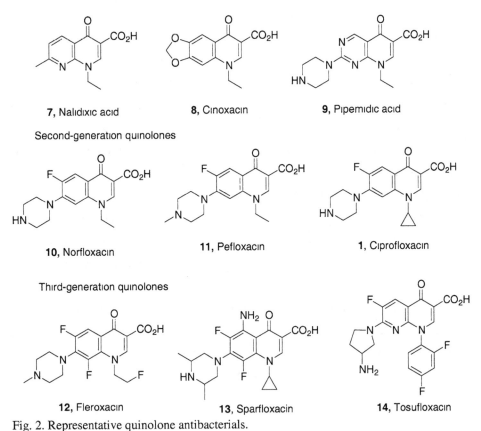

7, Nalidixic acid   8, Cinoxacin   9, Pipemidic acid

Second-generation quinolones

10, Norfloxacin   11, Pefloxacin   1, Ciprofloxacin

Third-generation quinolones

12, Fleroxacin   13, Sparfloxacin   14, Tosufloxacin

Fig. 2. Representative quinolone antibacterials.

Like all fluoroquinolone antibacterials, ciprofloxacin (1) targets bacterial DNA gyrase (a.k.a. DNA topoisomerase II) and DNA topoisomerase IV, an essential bacterial enzyme. DNA gyrase is a tetramer, composed of two A-subunits and two B-subunits, which catalyzes the supercoiling of cellular DNA by a nicking, pass-through, and re-sealing process. Ciprofloxacin (1) binds to the A-subunits and inhibits DNA gyrase, inducing cleavage of the DNA backbone, thus exerting the antibacterial effect. Ciprofloxacin (1) is one of the most potent fluoroquinolone antibacterials *in vitro*. It has good activity, conferred by the piperazinyl substituent at C_7, against Gram-negative pathogens (*Escherichia coli, Klebsiella pneumoniae, P. rettheri, Enterobacter cloacae,* and *Pseudomonas aeruginosa*) but is generally weaker against Gram-positive organisms. Ciprofloxacin (1) has MICs (**m**inimal **i**nhibition **c**oncentrations) against enterobacteriaceae ranging from 0.008 to 2.0 mg/L.

Anthrax is a bacterium that when inhaled, the spores travel to the lung and start disseminating and producing toxins, which can be lethal if left untreated. Ciprofloxacin (1) (Cipro®) is approved by FDA for the treatment of anthrax. Indeed, a 60-day regiment is effective to treat the inhaled form of anthrax after an individual has been exposed. However, it is not the only antibiotic to treat an anthrax infection. Tetracyclines such as deoxycycline and β-lactams such as penicillin work as well, although the older antibiotics are more prone to drug-resistance because they have been used for longer time. On the other hand, widespread use of Cipro® must be discouraged because it will accelerate the development of drug-resistance and obviates the effect.

In terms of pharmacokinetics, 70–80% of orally administered ciprofloxacin (1) is absorbed by the GI tract. It takes 0.5–1.0 h to reach a peak concentration, with a terminal half-life of 3.0–4.0 h. After oral administration, absorption was sufficient, and the absolute bioavailability varied between 63% to 77%.

Like all fluoroquinolone antibacterials, ciprofloxacin (1) causes articular damage in juvenile animals. Consequently, it is not recommended for children or pregnant women. Nonetheless, more data have emerged for its pediatric applications thanks to its high antibacterial effectiveness and convenience in oral administration.

§7.1.2 Synthesis of ciprofloxacin (1)[8–10]

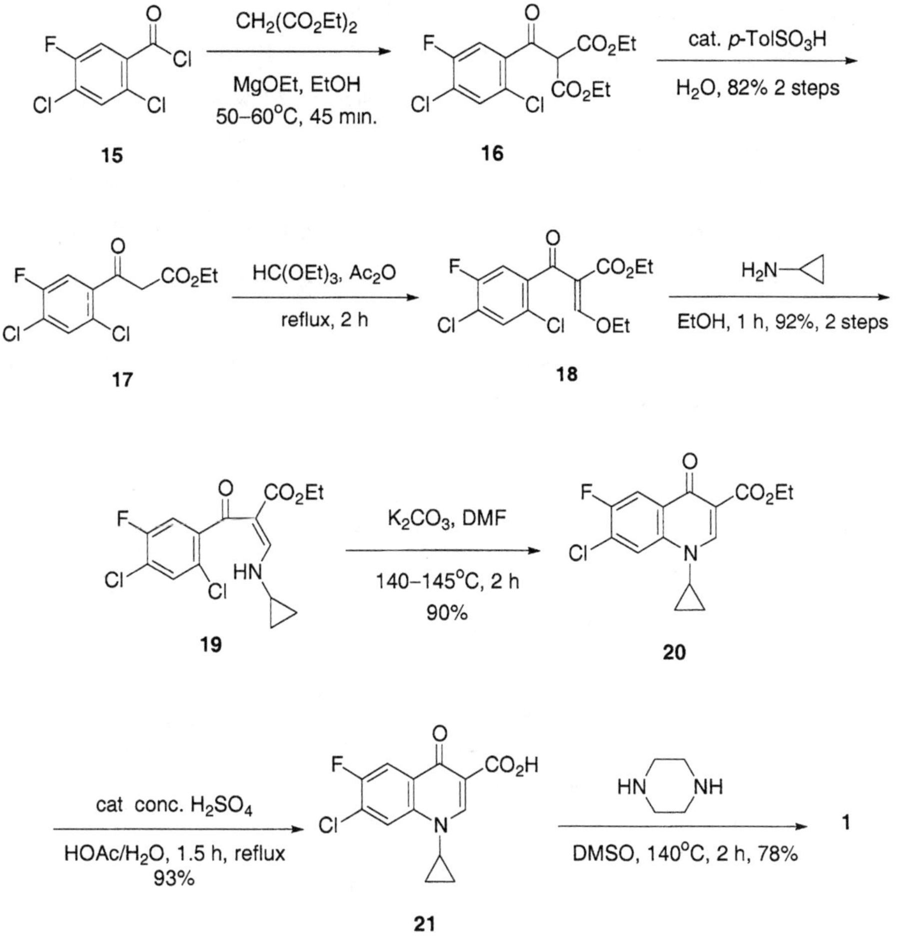

Scheme 1. The Bayer synthesis of ciprofloxacin (1).

The Bayer synthesis of ciprofloxacin (1) patented in 1981 utilized 2,4-dichloro-5-fluorobenzoyl chloride (15) as the starting material. With the aide of magnesium ethoxide, condensation of acid chloride 15 and diethyl malonate assembled ketone 16, which was subsequently decarboxylated using tosylic acid to form ethyl 2,4-dichloro-5-fluorobenzoylacetate (17) in 82% yield in two steps from 15. A Dieckman-like condensation of 17 with ethyl orthoformate was carried out in refluxing acetic anhydride to afford ethylacrylate **18**. When **18** was treated with cyclopropyl amine in ethanol, a Michael addition was followed by the subsequent expulsion of the ethoxy group to give enamine **19** with the stereochemistry as drawn. Under the influence of a base such as K_2CO_3, NaH or KH, an intramolecular S_NAr reaction of **19** took place to give the cyclized quinolone **20**. Hydrolysis of the ethyl ester on **20** was accomplished using catalytic concentrated sulfuric acid in a refluxing 1:1 acetic acid/H_2O mixture. Finally, a chemoselective S_NAr displacement of the resulting 6-fluro-7-chloroquinolone **21** with piperizine took place exclusively at the 7-chloro position to deliver ciprofloxacin (**1**). The chemoselectivity is a result of the activation effect of the carbonyl group at the *para*-position.

     This synthesis is a good example where advantages have been taken of the regio- and chemoselectivity of the haloarenes for S_NAr displacement reactions.

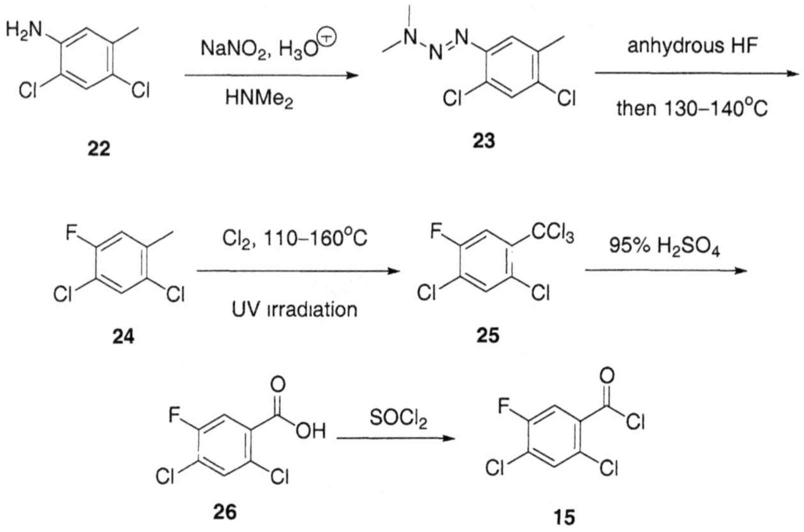

Scheme 2. The Bayer synthesis of 2,4-dichloro-5-fluorobenzoyl chloride (**15**).

On the other hand, 2,4-dichloro-5-fluorobenzoyl chloride (**15**) employed in Scheme 1 was prepared from a more readily available commercial source, 2,4-dichloro-5-methylphenylamine (**22**). Diazotization of aniline **22** and treatment with dimethylamine formed triazene **23**. When triazene **23** was dissolved in excess anhydrous HF, the corresponding 2,4-dichloro-5-methylphenyl diazonium fluoride was generated along with the release of dimethylamine. Without isolation, the aryl diazonium fluoride was heated at 130 to 140 °C to afford 3-fluorotoluene **24**. UV-initiated chlorination of **24** then gave rise to trichloromethylbenzene **25**, which was easily hydrolyzed with 95% sulfuric acid to form benzoic acid **26**. Acid **26** was then converted to the acid chloride **15** with thionyl chloride.

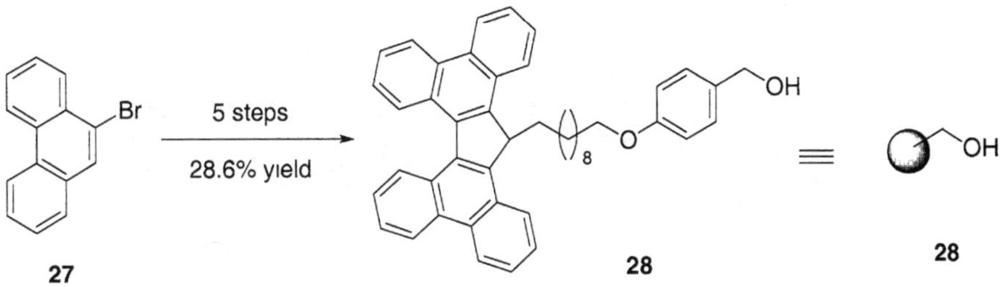

27   **28**   **28**

Scheme 3. Preparation of the anchor, tetrabenzo[*a,c,g,i*]-fluorene (**28**, Tbf).

A solid phase organic synthesis (SPOS) of ciprofloxacin (**1**) using a Wang resin was reported in 1996 by a group at Edinburgh.[11] More interestingly, the same group described an solid/solution phase synthesis of ciprofloxacin (**1**) using tetrabenzo[*a,c,g,i*]fluorene (**28**, Tbf) as an anchor.[12,13] The synthesis took advantage of the affinity of **28** with charcoal. Although all the reactions were carried out in solution phase, the purification was simply accomplished *via* absorption/desorption process mediated by switching between a polar solvent and a non-polar solvent. Thus, as illustrated in Scheme 3, tetrabenzo[*a,c,g,i*]-fluorene (**28**) was prepared from 9-bromophenanthrene (**27**) in 5 steps and 28.6% total yield. Meanwhile, β-keto-ester **31** was assembled from potassium ethyl malonate (**29**) and triflurobenzoyl chloride **30**. Trans-esterification of **31** with alcohol **28** was aided by DMAP in refluxing toluene to form the Tbf-keto ester **32**. Treatment of **32** with *N,N*-dimethylformamide diethyl acetal afforded the enamine **33**, which was converted to another enamine **34**. Due to the instability of enamines **33** and **34**, the reactions were carried out *in situ* and converted to quinolone **35** with the aide of tetramethylguanidine (TMG). The S_NAr of **35** with piperazine led to adduct **36**, which upon treatment with 90% aqueous TFA then released ciprofloxacin (**1**) in 57% yield.

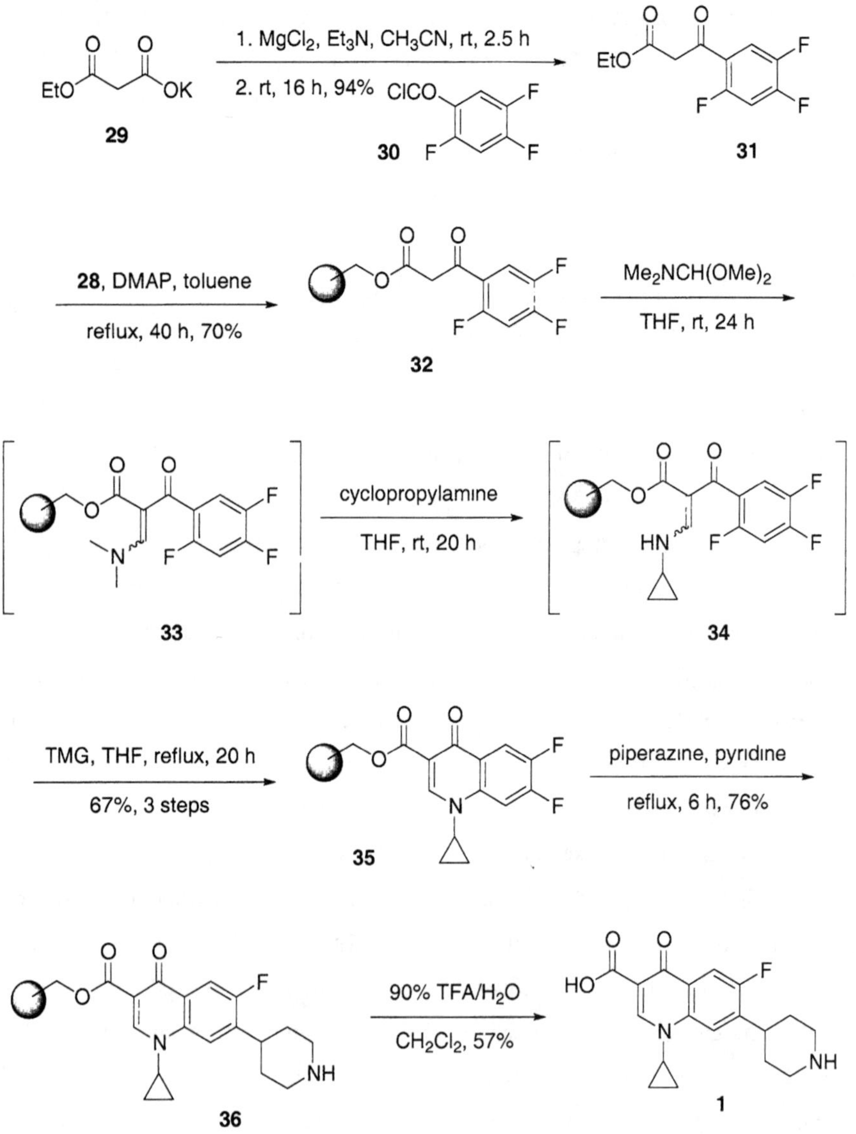

Scheme 4. The Edinburgh solid/solution phase organic synthesis of ciprofloxacin (**1**).

The aforementioned process has the advantage of being a homogeneous solution synthesis as well as the advantage of the heterogeneous purification.

§7.2   Linezolid (Zyvox®)

§7.2.1 Introduction to linezolid (2)[14–23]

The genesis of linezolid (2) began in 1978 when a DuPont patent described a novel oxazolidinone antibacterial agent S-6123 (37).   This compound and two subsequently optimized drug candidates DuP 721 and DuP 105 did not materialize as marketed drugs due to unacceptable toxicity. Inspired by this innovation, scientists at Upjohn further developed this class of compounds *via* intensive SAR studies, obtaining linezolid (2), a compound with favorable pharmacological, pharmacokinetic, and safety profiles.

Linezolid (2) is the first marketed member of a novel class of oxazolidinone antibacterial agents. Its mechanism of action (MOA) is *via* inhibition of the initial phase of bacterial protein synthesis. Due to this unique MOA, there has been no reported cross-resistance between oxazolidinones and other protein-synthesis inhibitors. Linezolid (2) has inhibitory activity against a broad range of Gram-positive bacteria, including methicillin-resistant *Staphylococcus aureus* (MRSA), glycopeptide-intermediate *S. aureus* (GISA), vancomycin-resistant enterococci (VRE) and penicillin-resistant *Streptococcus pneumoniae*. Linezolid (2) also shows activity against certain anaerobes, including *Clostridium perfringens, C. difficile, Peptostreptococcus* spp. And *Bacteroid fragilis.*

Linezolid (2) has a desirable pharmacokinetics profile.   Its mean absolute bioavailability is about 100% and its peak plasma concentrations (C_{max}) are achieved in 1 to 2 hours (t_{max}). The drug has a steady-state volume of distribution (V_{ss}) of 40–50 L and is moderately (31%) bound to plasma proteins. In addition, the tissue penetration of linezolid (2) is considered excellent in skin, soft tissue, lung, heart, intestine, liver, urine, kidney and cerebal synovial fluid (CSF).

In terms of metabolism, linezolid (2) is primarily metabolized by the oxidation of the morpholine ring, giving rise to two inactive metabolites, although the P450 enzyme system does not appear to be involved in the drug's metabolism. However, drug metabolism only accounts for less than 10% of the administered drug because the rest is recovered mostly in urine (80–85%) and feces (7–12%). The total clearance (CL) and renal clearance (CL_R) of linezolid (2) are 7.2 and 3 L/h (120 and 50 mL/min), respectively. Finally, the elimination half-life ($t_{1/2}$) is 4.5 to 5.5 hours at steady-state or after a single dose.

§7.2.2 Synthesis of linezolid (2)[24–29]

The Upjohn synthesis of linezolid (2) began with an S_NAr displacement of 3,4-difluoronitrobenzene (38) with morpholine. The 4-fluoro substituent was regioselectively displaced due to the activating effect of the *para* nitro substitution. The resulting adduct 39 was conveniently reduced to aniline 40 *via* hydrogenation. After protection of 40 with benzyl chloroformate, the resulting carbamate 41 became the precursor for installing the key oxazolin-2-one moiety. To this end, carbamate 41 was treated with n-BuLi, followed addition of (R)-glycidyl butyrate (42) to assemble the oxazolin-2-one 43. Mechanistically for this interesting reaction, the anionic nitrogen nucleophile presumably attacks the epoxide from the less hindered site to give an alkoxide. The resulting alkoxide then closes the ring to give the oxazolin-2-one with the pendant butyrate. Subsequently, the expelled benzyloxyl anion underwent a nucleophilic substitution of the butyrate to deliver oxazolin-2-one 43.

With oxazolin-2-one 43 in hand, simple functional group transformations ensued. Thus, mesylation of 43 using methylsufonyl chloride gave mesylate 44, which was followed by an S_N2 displacement with NaN_3 to produce azide 45. Reduction of the resulting azide **45** and subsequent acylation was accomplished in one pot to fashion linezolid (**2**).

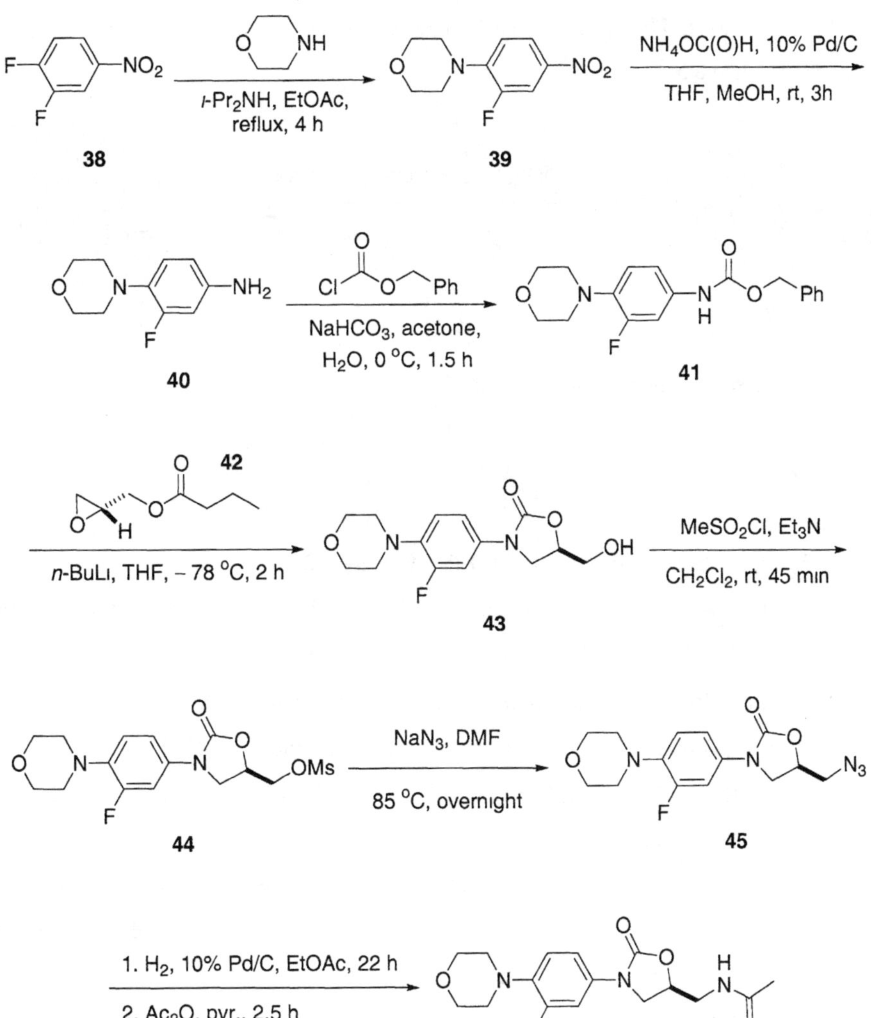

Scheme 5. The Upjohn synthesis of linezolid (**2**).

In a more proficient process chemistry synthesis and *in lieu* of (*R*)-glycidyl butyrate (**42**), transformation of carbamate **41** to linezolid (**2**) was also achieved by the using (S)-N-[2-(acetyloxy)-3-chloropropyl]acetamide (**46**).[29] Therefore, chloride **46** was prepared from the corresponding amino alcohol and acetic anhydride. Treatment of carbamate **41** with LiO*t*-Bu in DMF–MeOH was followed by addition of chloride **46** to give linezolid (**2**) in 72.1% yield. This route is five steps fewer than the original one depicted in Scheme 5.

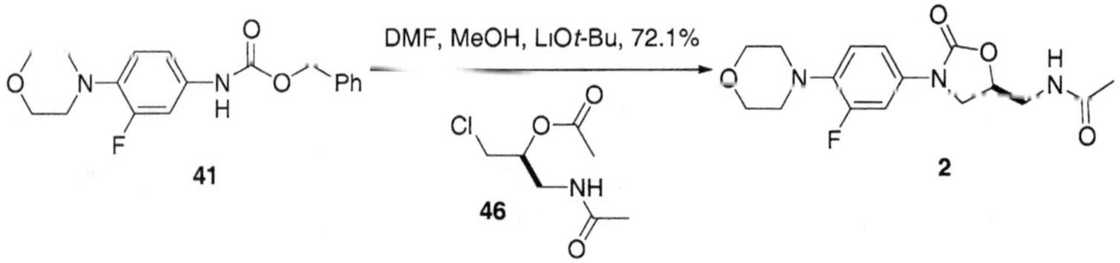

Scheme 6. An expedient route to linezolid (**2**).

§7.5   References

1.   Burnie, J.; Burnie, R. *Drugs Fut.* **1984**, *9*, 179–182.

2.   Höffken, G.; Lode, H.; Prinzing, C.; Borner, K.; Koeppe, P. *Antimicro. Ag. Chemother.* **1985**, *27*, 375–379.

3.   Watt, B.; Brown, F. V. *J. Antimicro. Chemother.* **1986**, *17*, 605–613.

4.   Watt, B.; Brown, F. V. *J. Antimicro. Chemother.* **1986**, *17*, 623–628.

5.   Scully, B. E.; Neu, H. C.; Parry, M. F.; Mandell, W. *Lancet* **1986**, *1*, 819–822.

6.   Kubin, R. *Infection* **1993**, *21*, 413–421.

7.   Chu, D. T. W.; Shen, L. L. In *Fifty Years of Antimicrobials: Past Perspective and Future Treads*, Hunter, P. A.; Darby, G. K.; Russell, N. J., Eds, Cambridge University Press: Cambridge, UK (1995).

8.   Grohe, K.; Zeiler, H.-J.; Metzger, K. DE 3142854 (**1983**).

9.   Grohe, K.; Zeiler, H.-J.; Metzger, K. US 4670444 (**1987**).

10.   Grohe, K.; Heitzer, H. *Libig Ann. Chim.* **1987**, 29–37.

11.   MacDonald, A. A.; Hobbs-Dewitt, S.; Hogan, E. M.; Ramage, R. *Tetrahedron Lett.* **1996**, *37*, 4815.

12.   Hay, A. M.; Hobbs-Dewitt, S.; MacDonald, A. A.; Ramage, R. *Tetrahedron Lett.* **1998**, *39*, 8721.

13.   Hay, A. M.; Hobbs-Dewitt, S.; MacDonald, A. A.; Ramage, R. *Synthesis* **1999**, 1979.

14.   Fugitt, R. B.; Luckenbaugh, R. W. US 4128654 (**1978**).

15.   Zyvox, Ford, C. W.; Hamel, J. C.; Wilsom, D. M.; Moerman, J. K.; Stapert, D.; Yancey, Jr., R. J.; Hutchinson, D. K.; Barbachyn, M. R.; Brickner, S. J. *Antimicro. Ag. Chemother.* **1996**, *40*, 1508–1513.

16.   Shinabarger, D. L.; Marotti, K. R.; Murray, R. W.; Lin, A. H.; Melchior, E. P.; Swaney, S. M.; Dunyak, D. S.; Demyan, W. F.; Buysse, J. M. *Antimicro. Ag. Chemother.* **1997**, *41*, 2132–2136.

17.   Dresser, L. D.; Rybak, M. J. *Pharmacotherapy* **1998**, *18*, 456–462.

18.   Lizondo, J.; Rabasseda, X.; Castañer, J. *Drugs Fut.* **1996**, *21*, 1116–1123.

19.   Diekema, D. J.; Jones, R. N. *Lancet* **2001**, *358*, 1975–1982.

20.   Xiong, Y.-Q.; Yeaman, M. R.; Bayer, A. S. *Drugs Today* **2000**, *36*, 631–639.

21.   Clemette, D.; Markham, A. *Drugs* **2000**, *59*, 815–827.

22.   Bouza, E.; Munoz, P.; *Clinical Microbiol. Infect.* **2001**, *7(Suppl. 4)*, 75–82.

23.   Perry, C. M.; Jarvis, B. *Drugs* **2001**, *61*, 525–551.

24.   Barbachyn, M. R.; Brickner, S. J.; Hutchingson, D. K. WO 95/07271 (**1995**).

25.   Brickner, S. J.; Hutchingson, D. K.; Barbachyn, M. R.; Manninen, P. R.; Ulanowicz, D. A.; Garmon, S. A.; Crega, K. C.; Hendges, S. K.; Toops, D. S.; Ford, C. W.; Zurenko, G. E. *J. Med. Chem.* **1996**, *39*, 673–679.

26.   Barbachyn, M. R.; Brickner, S. J.; Hutchingson, D. K. US 5688792 (**1997**).

27.   Pearlman, B. A.; Perrault, W. R.; Barbachyn, M. R.; Manninen, P. R.; Toops, D. S.; Houser, D. J. WO 97/37980 (**1997**).

28.   Pearlman, B. A. WO 99/24393 (**1999**).

29.   Perrault, W. R.; Pearlman, B. A.; Godrej, D. B. WO 02/085849 (**2002**).

Chapter 8. Atypical Antipsychotics

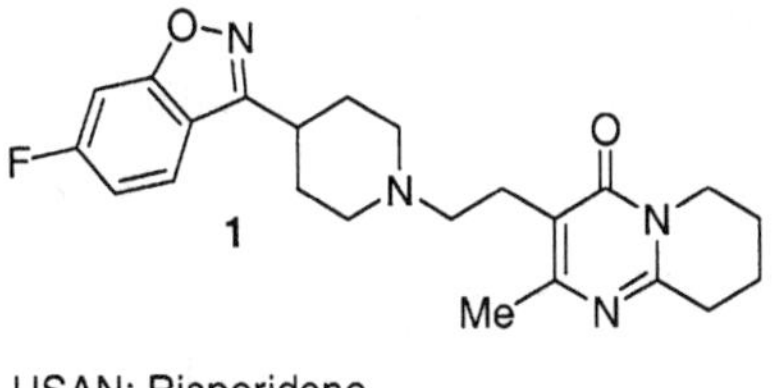

1

USAN: Risperidone
Trade Name Risperdal®
Company: Janssen
Launched: 1993
M W. 410.48

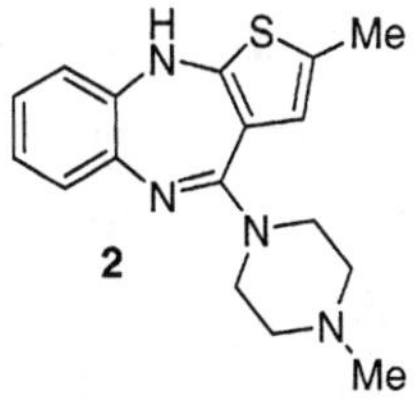

2

USAN: Olanzapine
Trade Name. Zyprexa®
Company: Eli Lilly
Launched: 1996
M.W. 312.43

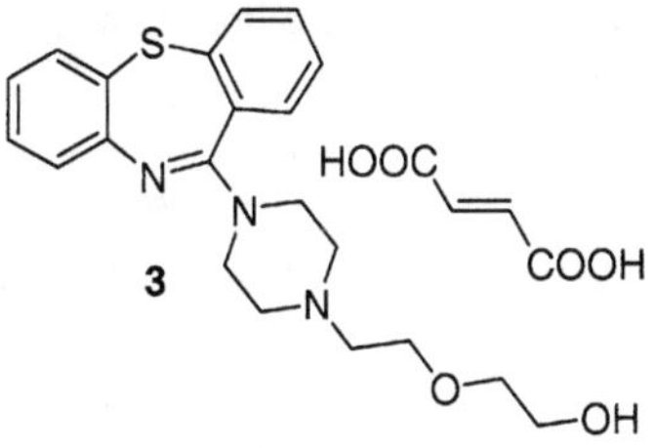

3

USAN: Quetiapine Fumarate
Trade Name: Seroquel®
Company. AstraZeneca
Launched: 1997
M.W. 383.51

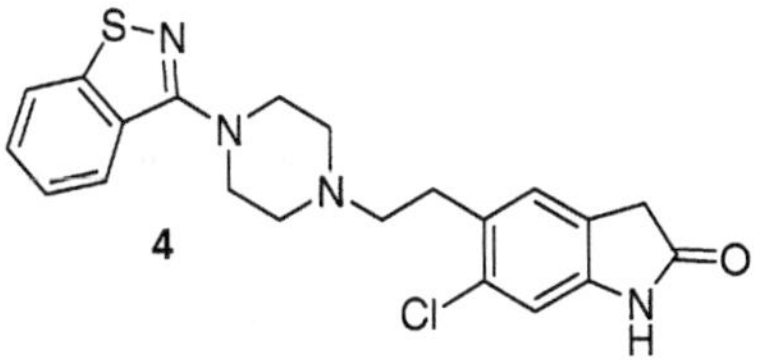

4

USAN: Ziprasidone
Trade Name: Geodon®
Company: Pfizer
Launched: 2001
M W. 412.94

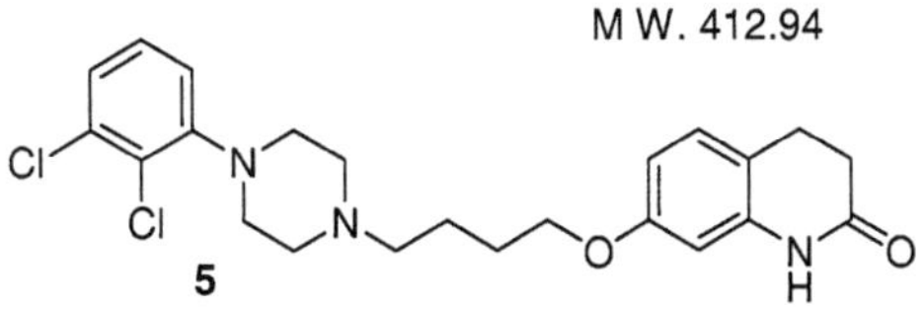

5

USAN: Aripiprazole
Trade Name Abilify®
Company: Bristol-Myers Squibb/Otsuka
Launched. 2002
M W. 448.38

§8.1   Background[1]

Schizophrenia is a mental disorder that is characterized by positive symptoms such as delusions, hallucinations and disorganized speech/behavior and negative symptoms including apathy, withdrawal, lack of pleasure and impaired attention.[1] Other symptom dimensions include depressive/anxious symptoms and aggressive symptoms such as hostility, verbal and physical abusiveness and impulsiveness.

The first conventional antipsychotic, chloropromazine (**6a**), was introduced in 1952. It was designed as an antihistamine, but was serendipitously discovered to possess antipsychotic properties. Subsequently, chloropromazine was shown to be a potent dopamine D_2 antagonist (K_i = 3 nM) with other pharmacologic properties that were thought to cause the unwanted side-effects. Haloperidol (**6b**) was developed as a more potent and selective D_2 antagonist. D_2-receptor blockade in the mesolimbic pathway is believed to reduce the positive symptoms of schizophrenia. Indeed, haloperidol is quite effective against the positive systems; however, it is ineffective in treating the negative symptoms and neurocognitive deficits of schizophrenia. In addition, administration of the drug typically causes extrapyramidal side-effects (EPS) including Parkinsonian symptoms, akathisia, dyskinesia and dystonia. Thus, the D_2-receptor antagonism of the conventional antipsychotics mediates not only their therapeutic effects, but also some of their side-effects. With the discovery of the newer atypical antipsychotics, the older conventional antipsychotics are no longer used for first-line therapy, but can still be effective as second-line or add-on treatments.

6a, Chloropromazine   **6b**, Haloperidol

Clozapine (**7**) is considered the first atypical antipsychotic. Atypical antipsychotics, sometimes called serotonin-dopamine antagonists (SDAs), have reduced EPS compared with conventional antipsychotics and are also believed to reduce negative, cognitive and affective symptoms of schizophrenia more effectively. All atypical antipsychotics are potent antagonists of serotonin 5-HT_{2A} and dopamine D_2 receptors; however, they also act on many other receptors including multiple serotonin receptors (5-HT_{1A}, 5-$HT_{1B/1D}$, 5-HT_{2C}, 5-HT_3, 5-HT_6, 5-HT_7), the noradrenergic system (α_1 and α_2), the cholinergic system (M_1) and the histamine receptors (H_1). The challenge remains to determine which of these secondary pharmacologic properties may be synergistic leading to improved efficacy, and which are undesired and account for the side-effects. It is generally accepted that an atypical antipsychotic should combine a minimum of 5-HT$_{2A}$ antagonism with D_2 antagonism in order to provide increased efficacy with fewer side-effects.[2] Serotonin–dopamine antagonists, but not conventional antipsychotics (dopamine antagonist without 5-HT$_{2A}$ antagonism), increase dopamine release in the mesocortical pathway. This provides a possible explanation for the improved efficacy of atypical antipsychotics in the treatment of negative symptoms of schizophrenia. Furthermore, 5-HT$_{2A}$ antagonism in the nigrostriatal pathway is believed to reduce EPS and tardive dyskinesia because dopamine release from this pathway is regulated by serotonin. If serotonin is not present at its 5-HT$_{2A}$-receptor on the nigrostriatal dopaminergic neuron, then dopamine is released.

Clozapine (**7**) was removed from the market in 1975 because of a drug-associated agranulocytosis, a potentially fatal blood disorder that results in lowered white-cell counts, which occured in approximately 2–3% of patients. Additonal side-effects of clozapine therapy include sedation (H_1), weight gain (5-HT$_{2C}$) and orthostatic hypotension (α_1). Clozapine was reintroduced in 1990 and is now relegated as a second-line treatment with extensive monitoring of the patient's blood cell count. However, over the years it has demonstrated efficacy against treatment-resistant schizophrenia and some still consider it to be the gold standard for treatment-refractory patients.

7, Clozapine (Clozaril®)   **8**, Zotepine (Zoleptil®)   **9**, Sertindole (Serlect®)

Risperidone (**1**), olanzapine (**2**)[6], quetiapine (**3**)[7] and ziprasidone (**4**)[8] are currently considered as the four first-line therapeutics for psychosis and will be highlighted in detail in this chapter. The newest antipsychotic to make its way to the market is aripiprazole (**5**).[9] It has a slightly different mechanism of action from the atypicals in that it is a D_2 partial agonist rather than a full antagonist. Each of these drugs has a unique pharmacological and clinical profile; therefore, the clinician must balance the benefits and risk factors for each patient in determining which drug to prescribe.[3-5]

Risperidone (**1**) has high affinity for D_2, 5-HT_{2C} and α_1 receptors and a very high affinity for the 5-HT_{2A} receptor. Risperidone is the most likely of the atypical antipsychotics to cause prolactin increases, but has a lower weight gain liability than olanzapine or quetiapine. Risperidone has a relatively narrow therapeutic window since doses above 6 mg/day cause EPS in a dose-dependent manner.

Olanzapine (**2**) is a close analog of clozapine where one of the benzene rings of the tricyclic nucleus is replaced with a thiophene ring. Olanzapine has high affinity for the 5-HT_{2A}, 5-HT_{2C}, H_1 and M_1 receptors and moderate affinity for the D_2 and α_1 receptors. Olanzapine is associated with high levels of weight gain (second only to clozapine). Olanzapine also causes some EPS at higher doses.

Quetiapine (**3**) has the lowest affinity for the D_2 and 5-HT_{2A} receptors among the atypicals; therefore, relatively high doses are required for maximal efficacy. Quetiapine causes significant weight gain, but less than that of olanzapine. Other side-effects include sedation, dizziness and hypotension.

Ziprasidone (**4**) has high affinity for the D_2 receptor, but even higher affinity for 5-HT_{2A} and 5-HT_{2C} receptors. Unlike other atypical antipsychotics, ziprasidone also has potent 5-$HT_{1B/1D}$ antagonist and 5-HT_{1A} partial agonist activity, as well as moderate SRI/NRI activity. This receptor profile suggests that ziprasidone may be useful in relieving some of the depressive/anxious symptoms of schizophrenia. Ziprasidone has moderate affinity for the H_1 and α_1 receptors and negligible affinity for the M_1 receptor. Ziprasidone is more likely to increase the QTc interval than other atypical antipsychotics, but it appears to have the lowest liability for bodyweight gain.

Zotepine (**8**) and sertindole (**9**) also belong to this class of atypical antipsychotics; however, they are used less frequently. Zotepine was introduced in Japan in 1982 and was approved for use in the United Kingdom in 1998, however it is still not approved in the United States. It has been associated with an increased risk of drug-induced convulsive seizures as well as significant weight gain, which has limited its use. Sertindole was introduced by Abbott in 1996 and was shown to be efficacious for the treatment of the positive and negative symptoms of schizophrenia. However, sertindole has recently been withdrawn from the market because it causes significant prolongation of the QTc interval, which may lead to a ventricular arrhythmia known as *torsades des pointes*.

Aripiprazole (**5**) is a D_2 partial agonist with an intrinsic activity of approximately 30%.[9] Therefore, it acts as an agonist on pre-synaptic autoreceptors, which have a high receptor reserve, and as an antagonist on D_2 post-synaptic receptors, where significant levels of endogenous dopamine exist and there is no receptor reserve.[10] The intrinsic activity of 30% for aripiprazole prevents D_2 blockade from rising above 70%, which is above the 65% D_2 occupancy needed for a clinical response but below the 80% D_2 occupancy where EPS is observed. Consistent with this partial agonist mechanism, EPS was not observed with aripiprazole even when striatal D_2 receptor occupancy values where above 90%.[11] Aripiprazole can be considered atypical since it is also an antagonist at $5\text{-}HT_{2A}$ receptors. It is also a partial agonist at $5\text{-}HT_{1A}$ receptors which may provide some benefit against some of the negative symptoms of schizophrenia. Preliminary clinical studies have demonstrated that aripiprazole is well tolerated and does not significantly induce EPS, weight gain, QT prolongation or increase plasma prolactin levels. It remains to be seen how effective aripiprazole will be in larger patient populations.

Much remains to be discovered about the underlying pathophysiology of schizophrenia and there is still a great need for medicinal chemists to develop more selective drugs that are devoid of clinically limiting side-effects and also address the cognitive impairment symptoms.

§8.2 Synthesis of risperidone (1)[12–20]

The U.S. patent covering risperidone was issued to Janssen in 1989 and the disclosed synthesis is shown in Scheme 1. The right-hand fragment of risperidone was prepared by refluxing 2-aminopyridine with 2-acetylbutyrolactone in the presence of POCl₃. Subsequent hydrogenation in acetic acid provided tetrahydropyrido[1,2-α]pyrimidinone **10** along with some dechlorinated by-product. The synthesis of the left-hand arylpiperidine was initiated by Friedel–Crafts acylation of 1,3-difluorobenzene with the acid chloride **11** to give **12**. The acetyl group of **12** was hydrolyzed with aqueous HCl. The resulting benzoylpiperidine was treated with hydroxylamine in refluxing ethanol to give oxime **13**, which cyclized to 1,2-benzisoxazole **14** when subjected to refluxing aqueous KOH. The two fragments were joined by alkylation of piperidine **14** with alkyl chloride **10** using Na₂CO₃ and KI in DMF. The product was recrystallized from DMF/*i*-PrOH to give risperidone (**1**) in 46% yield.

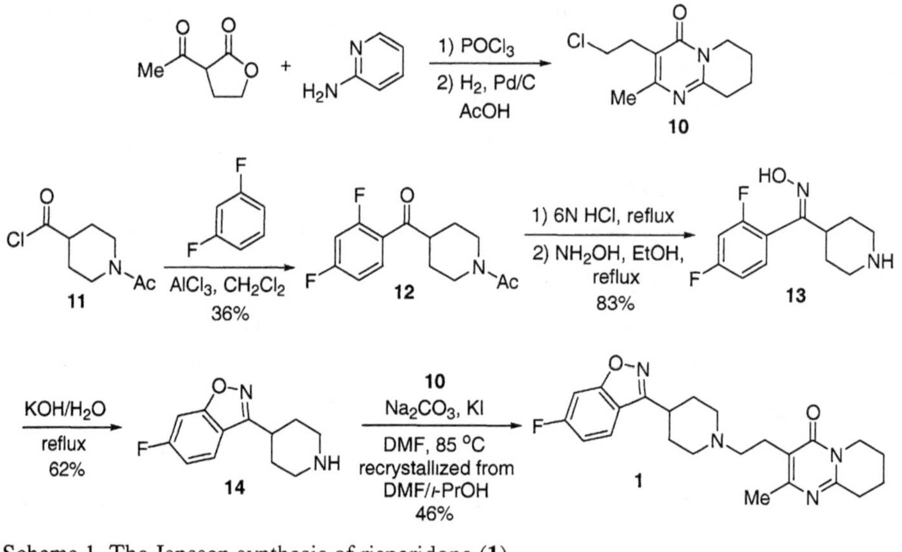

Scheme 1. The Janssen synthesis of risperidone (**1**).

A group from Spain patented a similar synthesis of risperidone (**1**) using similar synthetic transformations but in a different sequence (Scheme 2). Reaction of 2-aminopyridine with 2-acetylbutyrolactone in polyphosphoric acid at 160 °C afforded alcohol **15**, which was hydrogenated to **16** and then converted to chloride **10** with thionyl chloride. Benzoyl piperidine **19** was prepared in a similar manner as in Scheme 1. In contrast to Scheme 1, **19** was alkylated with **10** prior to conversion to the 1,2-

benzisoxazole to provide **20** in 63% yield. The benzisoxazole was formed in the final step to afford risperidone (**1**).

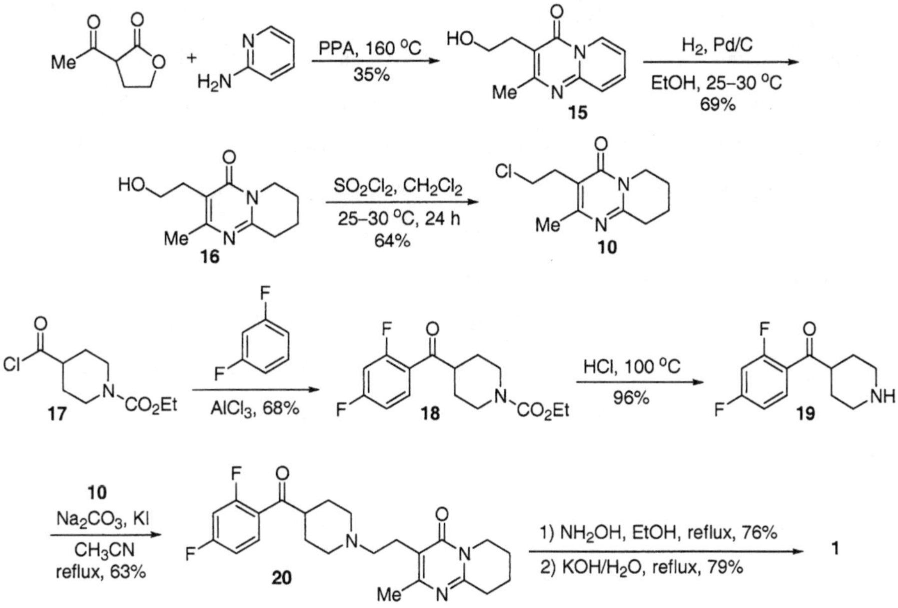

Scheme 2. The Spanish synthesis of risperidone (**1**).

Recently, a group from RPG Life Sciences in India reported a supposedly improved synthesis of risperidone (Scheme 3). Hydrogenation of **21** in aqueous HCl, promoted an efficient reduction without any dechlorination by-products. Furthermore, the subsequent alkylation of **14** with **10** was carried out under aqueous conditions in the presence of an inorganic base such as Na_2CO_3. It was hypothesized that the increased solubility of the inorganic base in the aqueous medium allows for more effective neutralization of the acid by-product (HCl), leading to less degradation of risperidone (**1**) resulting in higher yields and purity.

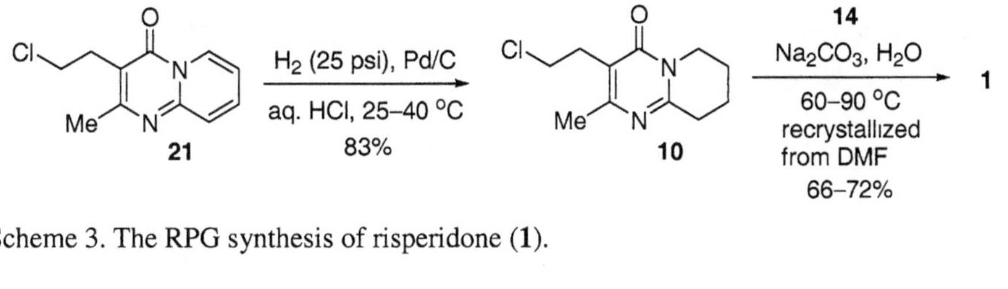

Scheme 3. The RPG synthesis of risperidone (**1**).

Teva Pharmaceutical Industries recently published a patent claiming the preparation of several novel crystal forms (polymorphs) of risperidone (**1**), namely, forms A, B and E (Scheme 4). The crystal form can have a large impact on the pharmaceutical properties of a drug. For instance, the solubility of the different polymorphs may vary dramatically. Compound **1** was prepared by refluxing **10** and **14** in isopropanol in the presence of Na_2CO_3 and KI for 9 hours. Recrystallization from isopropanol or acetone provided high-purity risperidone form A in 60–63% yield. These reaction and recrystallization conditions avoid the use of DMF, which is difficult to remove. Risperidone form B could be formed by recrystallization from $CHCl_3$/cyclohexane or by dissolving **1** in aqueous HCl and adding aqueous Na_2CO_3 to facilitate precipitation. Form E could be prepared by dissolving **1** in isopropanol and then adding water to facilitate precipitation.

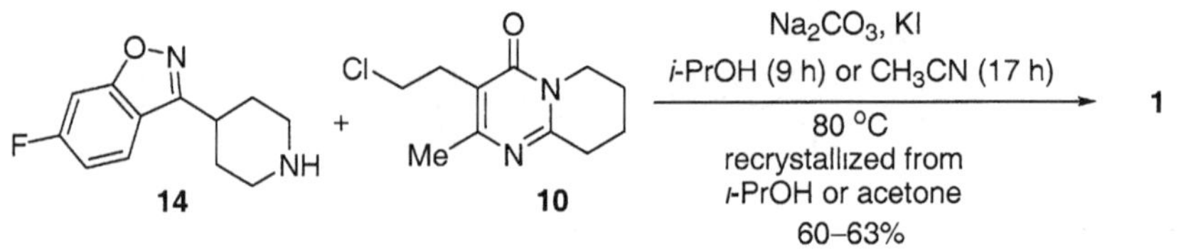

Scheme 4. The Teva synthesis of risperidone (**1**).

§8.3   Synthesis of olanzapine (2)[21–28]

Chakrabarti and coworkers at Eli Lilly in the United Kingdom have reported the initial discovery and synthesis of olanzapine (Schemes 5 and 6). The thiophene **22** was synthesized by adding a DMF solution of malononitrile to a mixture of sulfur, propionaldehyde and triethylamine in DMF. The anion of amino thiophene **22** underwent a nucleophilic aromatic substitution with 2-fluoronitrobenzene to provide **23**. The nitro group was reduced with stannous chloride and the resulting aniline cyclized with the cyano group to form amidine **24**. Finally, a mixture of N-methylpiperazine and **24** were refluxed in DMSO/toluene to afford olanzapine (**2**).

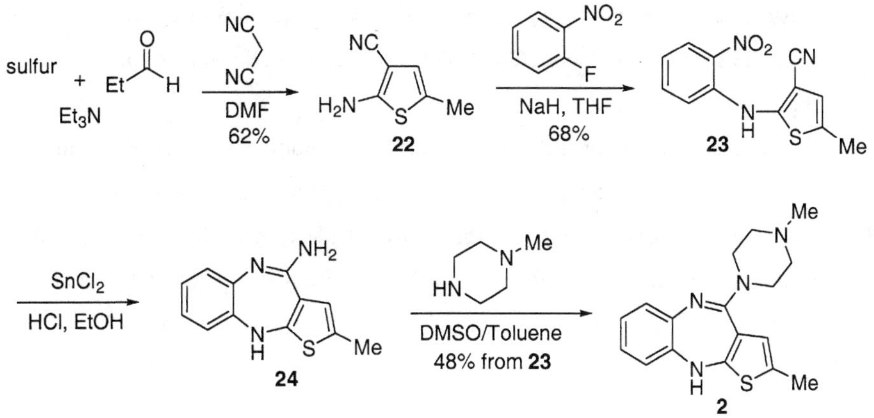

Scheme 5. The Eli Lilly synthesis of olanzapine (2).

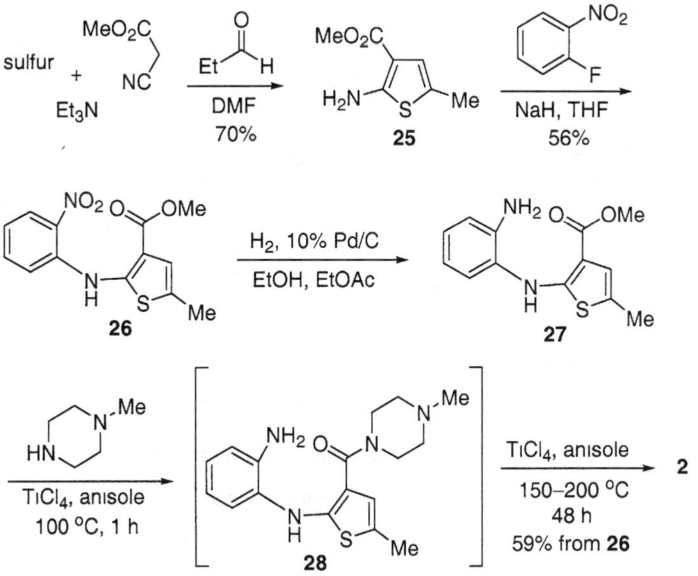

Scheme 6. An alternate synthesis of olanzapine (2).

Alternatively, substituting malononitrile with methyl cyanoacetate in the thiophene forming reaction gave **25** with a carbomethoxy group at the 3 position of the thiophene (Scheme 6). Compound **25** was reacted with 2-fluoronitrobenzene as in Scheme 5 to form **26** and hydrogenation of the nitro group provided **27**. The crude diamino ester was reacted with *N*-methylpiperazine in the presence of TiCl$_4$ at 100 °C for 1 h to give the intermediate amide **28**, which was heated under reflux for 48 hours to effect ring closure to deliver **2**.

Cen at the Shanghai Institute of Pharmaceutical Industry has recently published a synthesis of olanzapine (Scheme 7). The thiophene **22** was treated with 2-chloronitrobenzene in the presence of lithium hydroxide to give **23**. Reduction of the nitro group and subsequent ring closure gave **24**. Addition of piperazine to the amidine **24** followed by methylation provided olanzapine (**2**) in an overall yield of 29%.

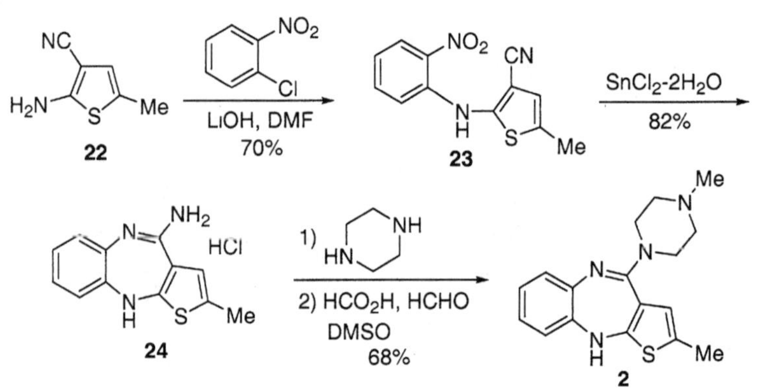

Scheme 7. The Shanghai Institute synthesis of olanzapine (**2**).

Several patents have recently published that claim the preparation of several different hydrates and polymorphic crystal forms of olanzapine. Dr. Reddy's Laboratories has recently disclosed the preparation of the monohydrate and the dihydrate of olanzapine. A mixture of **24** and *N*-methylpiperazine was refluxed in DMSO and toluene and then cooled (Scheme 8). Water was added and the precipitate was filtered and washed with water. The resulting solid was placed under vacuum at 30 to 50 °C to give the monohydrate or at ambient temperature to give the dihydrate. Recrystallization of crude olanzapine or one of its hydrates from CH$_2$Cl$_2$ provided crystal form 1, whereas recrystallization from EtOAc provides crystal form 2.

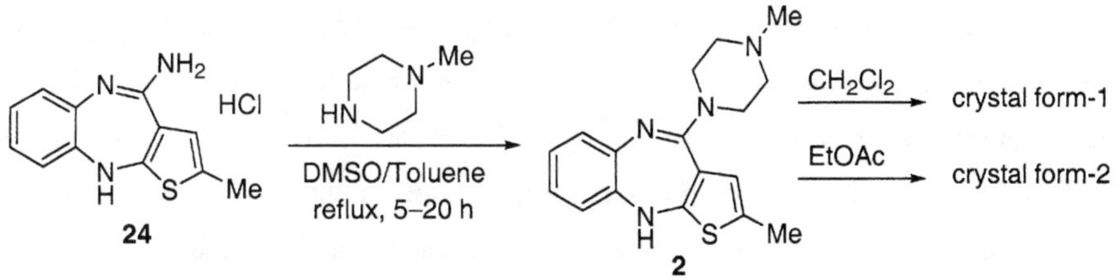

Scheme 8. Dr. Reddy's synthesis of olanzapine (2).

§8.4   Synthesis of quetiapine fumarate (3)[29–34]

The initial synthesis of quetiapine was disclosed by Warawa and Migler in a patent issued to ICI (now AstraZeneca) in 1987 (Scheme 9). The dihydrodibenzothiazepinone 31 was prepared according to the method of Schmutz. Displacement of the chloro group of o-chloronitrobenzene with thiophenoxide followed by hydrogenation of the nitro group using Raney-Nickel gave aniline 29. Isocyanate 30 was formed by treating 29 with phosgene in toluene. Ring closure to the core tricyclic structure 31 was effected by refluxing 30 in H_2SO_4. Imino chloride 32 was formed by treating 31 with phosphorous oxychloride in the presence of dimethylaniline. A mixture of 32 and 1-(2-hydroxyethoxy)ethylpiperazine was refluxed in xylenes to afford quetiapine which was converted to the fumarate salt 3.

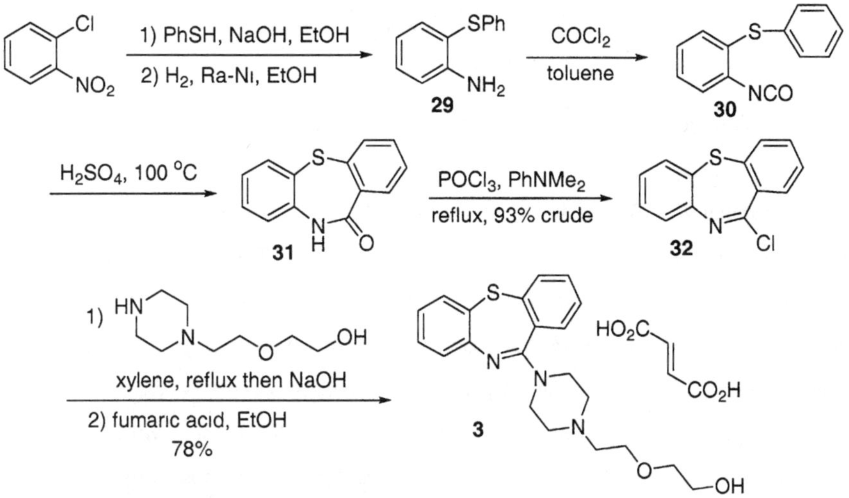

Scheme 9. The initial AstraZeneca synthesis of quetiapine fumarate (3).

Barker and Copeland have reported a slightly modified process for the preparation of **3** (Scheme 10). **29** was reacted with phenylchloroformate to give carbamate **33**. The ring closure was affected by heating **33** in polyphosphoric acid at 100 °C to provide **31**. The imino chloride **32**, formed as in Scheme 9, was treated with piperazine in refluxing toluene to give **34**. The piperazine **34** was then alkylated with 2-chloroethoxyethanol in the presence of NaI and Na$_2$CO$_3$ to afford quetiapine, which was isolated as the fumarate salt **3**.

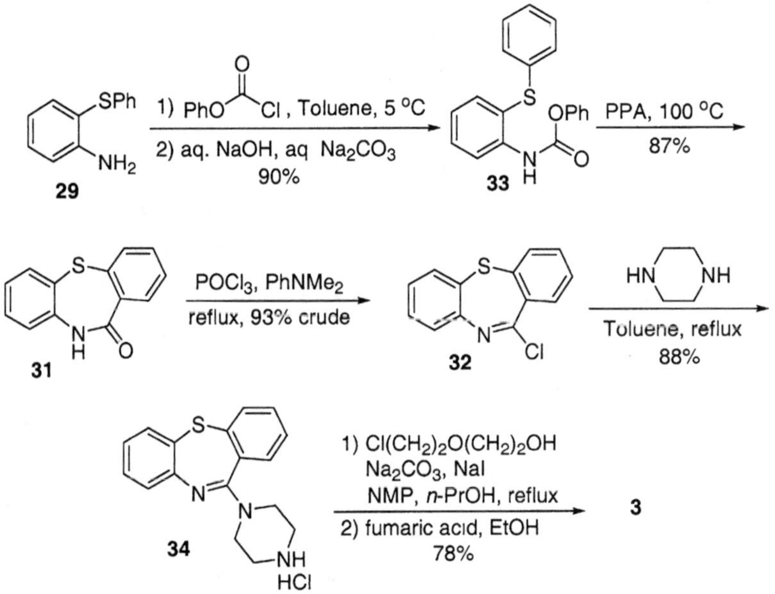

Scheme 10. The second AstraZeneca synthesis of quetiapine fumarate (**3**).

A company in Budapest recently reported an alternative synthesis of quetiapine aimed to provide a more economical process (Scheme 11). They sought to avoid the use of imino chloride **32** as an intermediate, because it is rather unstable and is easily hydrolyzed. The carbamate **33** was reacted with 1-(2-hydroxyethyl)piperazine to give crystalline **35** in 95% yield. The hydroxyethylpiperazine **35** was refluxed in thionyl chloride to provide crystalline chloroethylpiperazine **36** as the hydrochloride salt. Compound **36** was treated with phosphorous oxychloride and phosphorus pentoxide at reflux and **37** was isolated as a crystalline solid from diisopropyl ether in 75% yield.

Sodium metal was dissolved in ethylene glycol and a solution of **37** in toluene was added. The mixture was heated at 100 °C and a 98% yield of the free base **3** was obtained. The free base was treated with fumaric acid in ethanol and **3** was isolated as the crystalline fumarate salt in 85% yield.

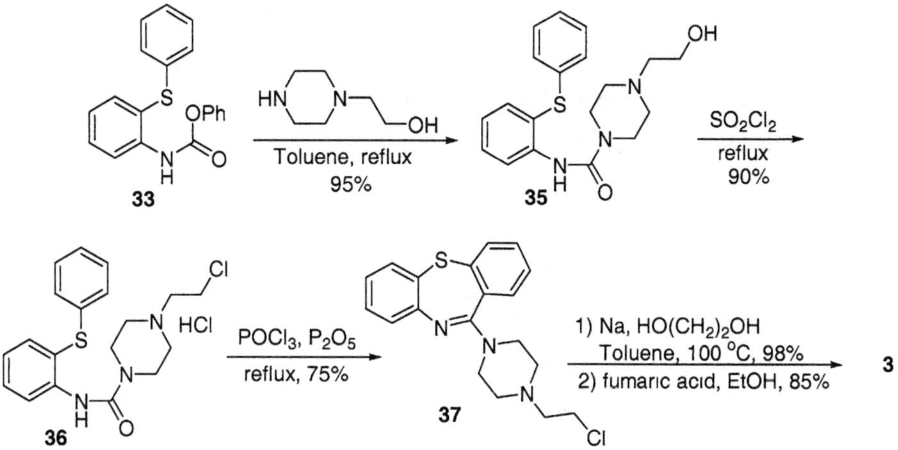

Scheme 11. The Budapest synthesis of quetiapine fumarate (**3**).

§8.5   Synthesis of ziprasidone (4)[35–42]

The first synthesis of ziprasidone was disclosed by Lowe and Nagel in a patent issued to Pfizer in 1989. Initially, the left-hand fragment of ziprasidone, 3-benzisothiazolylpiperazine (**40**), was prepared according to the method of Yevich and coworkers (Scheme 12). The amide **38** was treated with phosphorous oxychloride to give 3-chloro-1,2-benzisothiazole (**39**). Compound **39** was then reacted with molten piperazine at 125 °C to provide piperazine **40** in 68% yield.

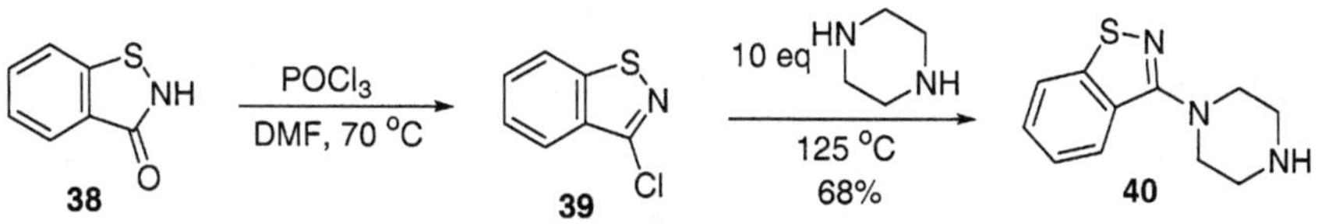

Scheme 12. Synthesis of 3-(1,2-benzisothiazolyl)piperazine (**40**).

The synthetic route in Scheme 12 was not suitable for pilot plant scale-up so the Pfizer process group improved the synthesis as shown in Scheme 13. Reaction of disulfide **41** with excess anhydrous piperazine in the presence of DMSO (2.2 equiv) and a small amount of isopropanol at 120 °C for 24 hours afforded piperazine **40** in 80% yield. DMSO was added to reoxidize the liberated 2-mercaptobenzonitrile back to the starting disulfide **41**, therefore product yields include utilization of both halves of the symmetrical disulfide **41**.   The reaction proceeds through trapping of the initial sulfenamide intermediate with an additional equivalent of piperazine to give the benzamidine intermediate **42**. Intermediate **42** then cyclizes with elimination of piperazine to give 3-(1,2-benzisothiazolyl)piperazine (**40**)

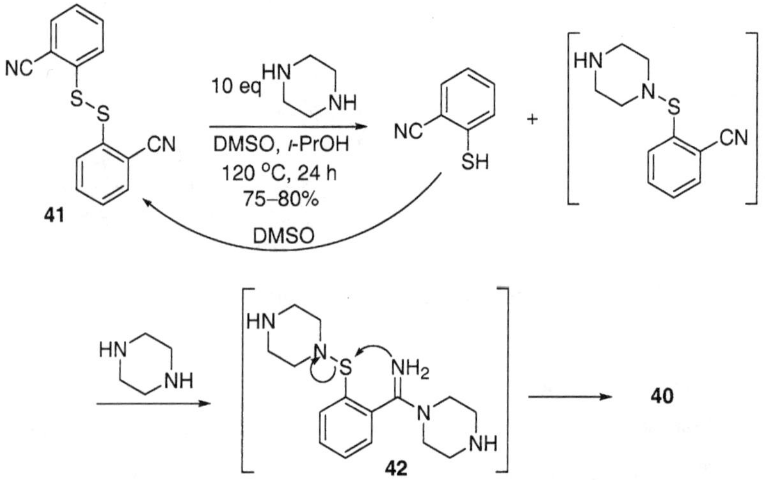

Scheme 13. The Pfizer process synthesis of 3-(1,2-benzisothiazolyl)piperazine (**40**).

The synthesis of the right-hand fragment of ziprasidone started with a Wolff–Kishner reduction of isatin **43** to give the oxindole **44** (Scheme 14). Friedel–Crafts acylation with chloroacetyl chloride afforded aryl ketone **45**, which was reduced with triethylsilane in trifluoroacetic acid to the phenethyl chloride **46**. The two fragments were joined by alkylation of **40** with **46** in the presence of NaI and Na$_2$CO$_3$ to give ziprasidone (**4**) in low yield. The yield of the coupling step was improved dramatically when the reaction was conducted in water (Scheme 15).

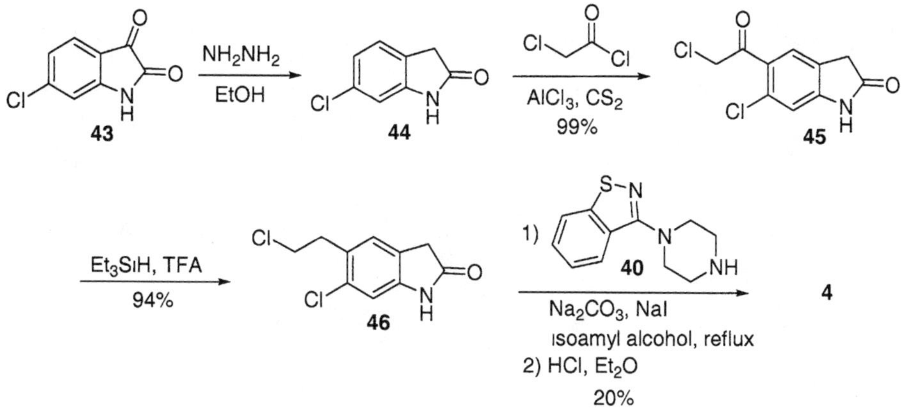

Scheme 14. Synthesis of ziprasidone (**4**).

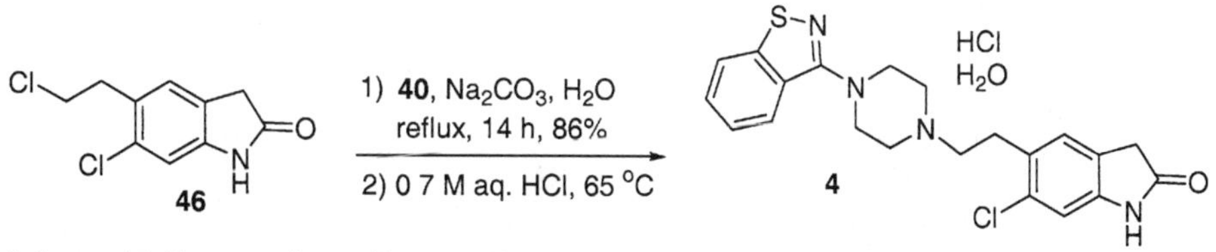

Scheme 15. Improved coupling conditions for the synthesis of ziprasidone (**4**).

The Pfizer process group has also developed two alternative syntheses of ziprasidone (Schemes 16 and 17). The first synthesis began with the nitration of 2,5-dichlorotoluene to give **48** in 56% yield. Compound **48** was refluxed with *t*-butoxy-bis(dimethylamino)methane to give the enamine **49**. Piperazine **40** was reacted with enamine **49** in acetic acid to afford the coupled piperazinyl enamine which was reduced with sodium triacetoxyborohydride to provide **50**. Dimethymalonate was added to the *o*-chloronitrobenzene **50** in the presence of potassium hydroxide to give **51** in moderate yield. Alternatively, use of the more acidic reagent methyl cyanoacetate under the same conditions resulted in a 82% yield of the adduct, which was treated with hydrochloric acid to give **52**. Malonate **51** was hydrolyzed and decarboxylated in aqueous hydrochloric acid to provide **52**. The acid **52** was esterified and the nitro group was reduced with sodium hydrosulfite with concomitant ring closure to provide ziprasidone (**4**).

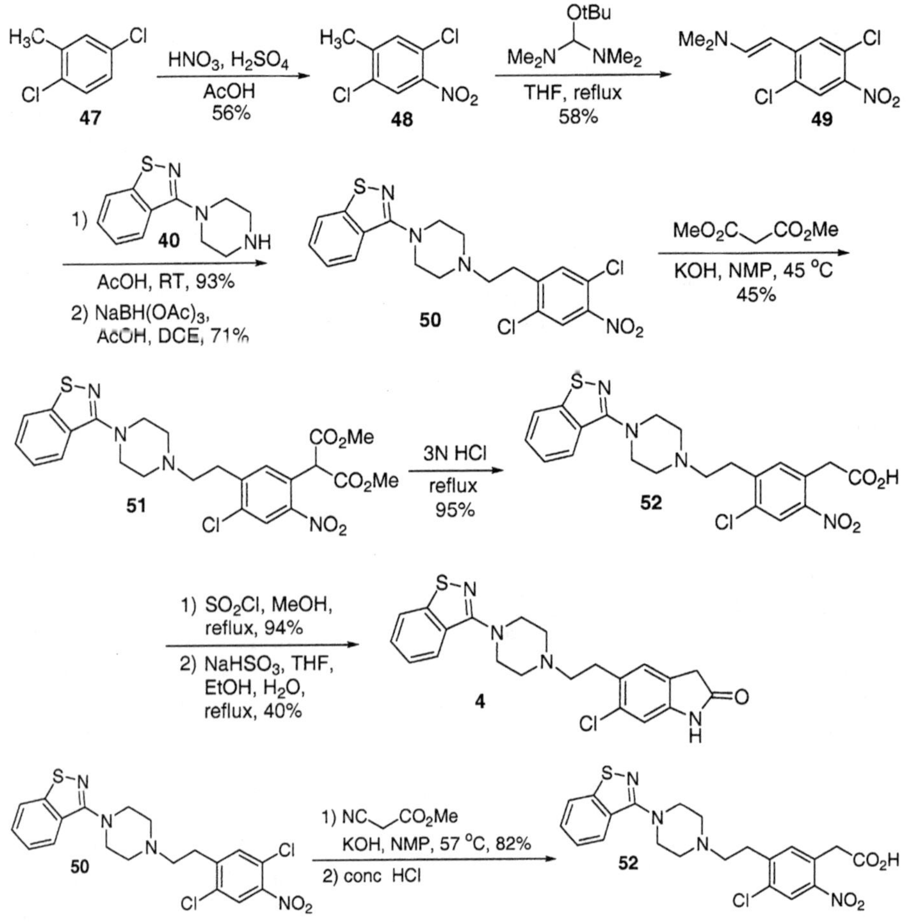

Scheme 16. An alternate Pfizer process synthesis of ziprasidone (4).

A second alternative synthesis of **4** is highlighted in Scheme 17. Enamine **49** was treated with aqueous oxalic acid to give the imminium salt, which was readily hydrolyzed under the reaction conditions to give aldehyde **53**. Reductive amination of **53** with Boc-piperazine afforded **54** in excellent yield. The sodium enolate of dimethyl malonate was added to **54** to give **55** in 48% yield. Refluxing **55** in hydrochloric acid resulted in removal of the Boc protecting group and hydrolysis of the methyl esters followed by a decarboxylation to give the mono-acid, which was converted back to the methyl ester (**56**) with thionyl chloride in methanol. Piperazine **56** was added to triflate **58**, prepared from **38**, in the presence of triethylamine to afford **59** in low yield. The nitro group was then reduced with sodium hydrosulfite, which also resulted in ring closure as in Scheme 16 to provide ziprasidone (**4**).

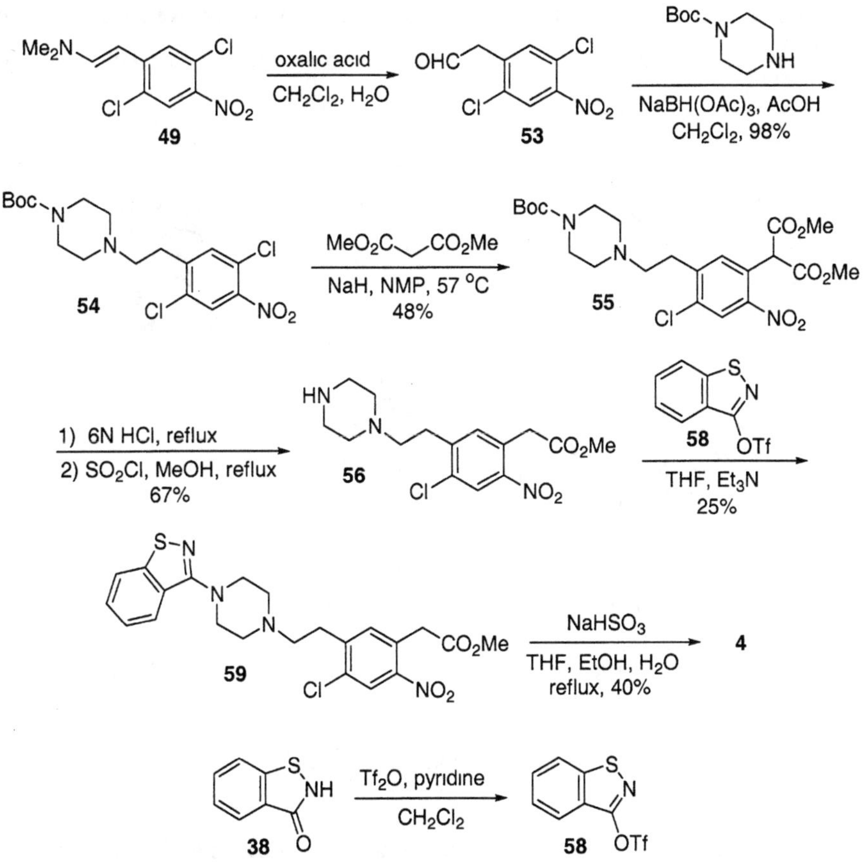

Scheme 17. Another Pfizer process synthesis of ziprasidone (**4**).

Pfizer has also prepared both ^{3}H- and ^{14}C-labelled ziprasidone (Schemes 18 and 19) to determine its metabolism and tissue distribution. It was envisioned that tritium could be introduced in the last step of the synthesis utilizing the selective replacement of a bromine atom on the benzisothiazole ring. Therefore, the synthesis began with the bromination of **39** using bromine in acetic acid with FeCl$_3$ catalysis. The dibrominated regioisomer **60** was isolated by liquid chromatography in 18% yield and reacted with piperazine in refluxing diglyme to give **61**. Alkylation of piperazine **61** with the alkyl chloride **46** from Scheme 14 in aqueous Na_2CO_3 afforded brominated ziprasidone (**62**) in excellent yield. **62** was treated with tritium gas using 10% Pd/BaSO$_4$ in THF for 6 hours resulting in a 53% conversion to **63**, which was isolated by preparative HPLC. Further exposure to the reaction conditions led to some dechlorination of the oxindole portion of the molecule.

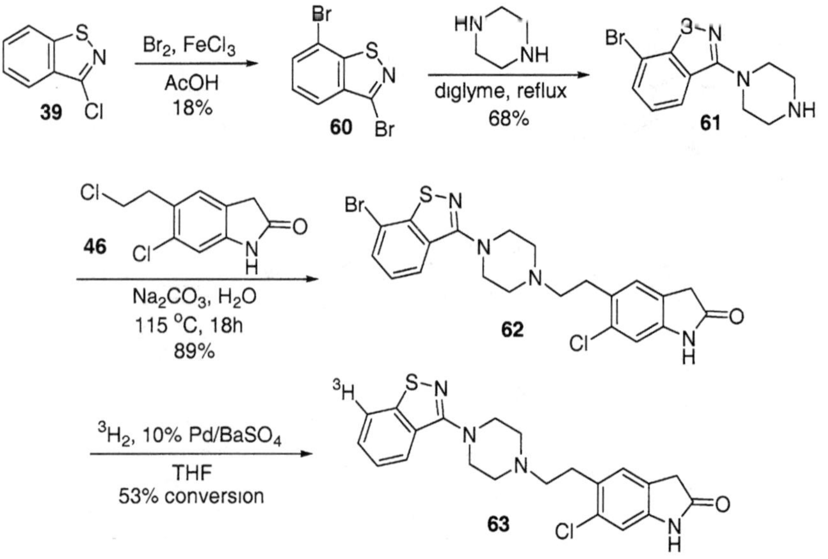

Scheme 18. Synthesis of tritium-labelled ziprasidone (**63**).

Preparation of the ^{14}C-labelled compound **65** was accomplished in a manner analogous to Scheme 14. Friedel–Crafts acylation of **44** was conducted with [2-^{14}C]-chloroacetyl chloride under aluminum trichloride catalysis to give the radiolabelled intermediate **64** (48 mCi/mmol). The carbonyl group of **64** was reduced with triethylsilane and the resulting alkyl chloride was reacted with piperazine **40** to provide ^{14}C-labelled ziprasidone (**65**). The HCl salt of **65** was formed resulting in a final compound with a specific activity of 9.6 mCi/mmol.

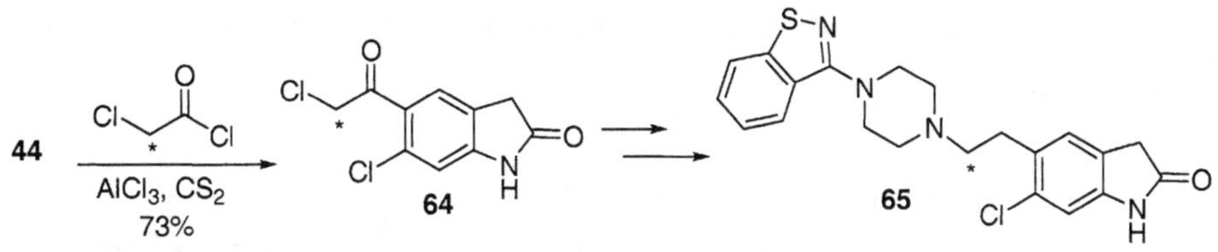

Scheme 19. Synthesis of ^{14}C-labelled ziprasidone (65).

§8.6   Synthesis of aripiprazole (5)[43-46]

Bristol–Myers Squibb in partnership with Otsuka has recently marketed aripiprazole for the treatment of schizophrenia. The synthesis (Scheme 20) begins with acylation of 3-methoxyaniline followed by Friedel–Crafts ring closure to give quinolinone 67. Hydrogenation provides dihydroquinolinone 68, which is treated with 1,4-dibromobutane in the presence of K_2CO_3 to afford 69. Compound 69 was treated with NaI and then alkylated with 2,3-dichlorophenylpiperazine to give aripiprazole (5).

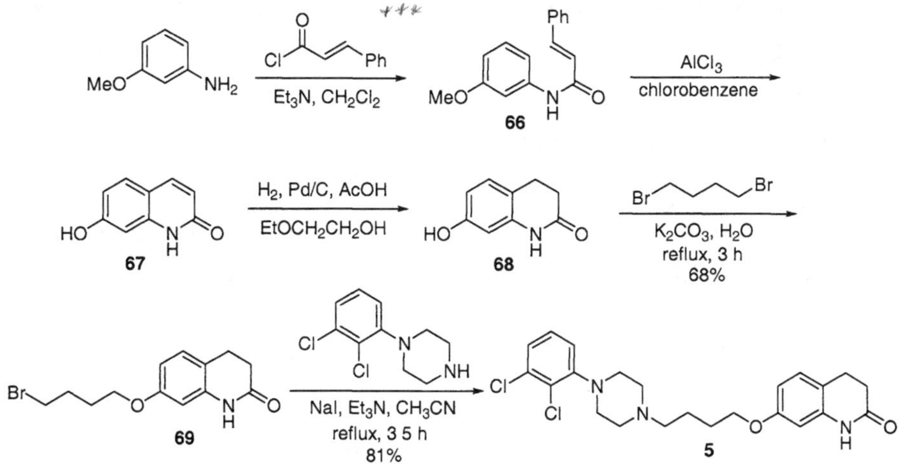

Scheme 20. The Otsuka synthesis of aripiprazole (5).

The 2,3-dichloro-4-hydroxyphenyl derivative of aripiprazole was prepared to confirm the structure of a primary metabolite of aripiprazole (Scheme 21). The synthesis began with the protection of 4-bromo-2,3-dichlorophenol as its benzyl ether 71. Palladium-catalyzed amination of 71 with piperazine proceeded regioselectively in excellent yield. Alkylation of the piperazine 72 with 69 in the presence of K_2CO_3 and

NaI, followed by removal of the benzyl group with concentrated HCl in acetic acid afforded the aripiprazole metabolite **73**.

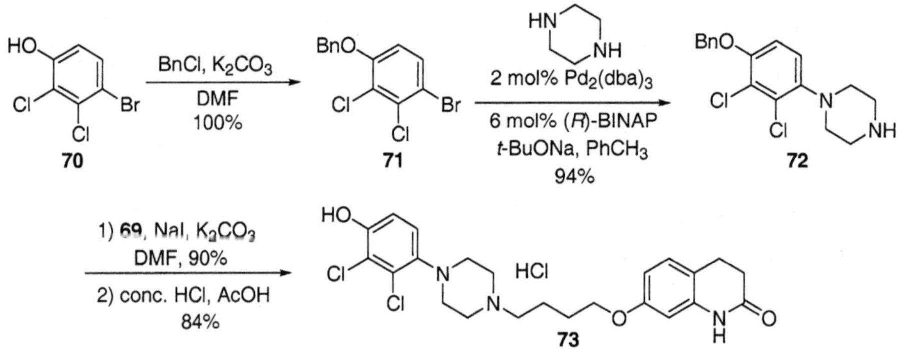

Scheme 21. Synthesis of the 4-hydroxyphenyl metabolite of aripiprazole.

§8.7   References

1.   Rowley, M.; Bristow, L. J.; Hutson, P. H. *J. Med. Chem.* **2001**, *44*, 477. Capuano, B.; Crosby, I. T.; Lloyd, E. J. *Curr. Med. Chem.* **2002**, *9*, 521. Kapur, S.; Remington, G. *Annu. Rev. Med.* **2001**, *52*, 503. Kelleher, J. P.; Centorrino, F.; Albert, M. J.; Baldessarini, R. J. *CNS Drugs*, **2002**, *16*, 249.

2.   Meltzer, H. Y. *Neuropsychopharmacology*, **1999**, *21*, 106S.

3.   (Weight gain) Casey, D. E.; Zorn, S. H. *J. Clin. Psychiatry*, **2001**, *62* (suppl 7), 4.

4.   (QTc) Goodnick, P. J.; Jerry, J.; Parra, F. *Expert Opin. Pharmacother.* **2002**, *3*, 479.

5.   (Prolactin) Goodnick, P. J.; Rodriguez, L.; Santana, O. *Expert Opin. Pharmacother.* **2002**, *3*, 13814.

6.   (Olanzapine) Lund, B. C.; Perry, P. J. *Exp. Opin. Pharmacother.* **2000**, *1*, 305. Kando, J. C.; Shepski, J. C.; Satterlee, W.; Patel, J. K.; Reams, S. G.; Green, A. I. *Ann. Pharmacother.* **1997**, *31*, 1325.

7.   (Quetiapine) Goldstein, J. M. *Drugs of Today*, **1999**, *35*, 193.

8.   (Ziprasidone) Buckley, P. F. *Drugs of Today*, **2000**, *36*, 583. Gunasekara, N. S.; Spencer, C. M.; Keating, G. M. *Drugs*, **2002**, *62*, 1217. Remington, G. *Expert. Rev. Neurotherapeutics*, **2002**, *2*, 13. Caley, C. F.; Cooper, C. K. *Ann. Pharmacother.* **2002**, *36*, 839.

9. (Arıpıprazole) Goodnick, P. J.; Jerry, J. M. *Expert Opin. Pharmacother.* **2002**, *3*, 1773. Ozdemir, V.; Fourie, J.; Ozdener, F. *Current Opinion in Investigational Drugs*, **2002**, *3*, 113. McGavin, J. K.; Goa, K. L. *CNS Drugs*, **2002**, *16*, 779. Taylor, D. M. *Int. J. Clin. Pract.* **2003**, *57*, 49.

10. Stahl, S. M. *J. Clin. Psychiatry*, **2001**, *62*, 841 and 923. Tamminga, C. A.; Carlsson, A. *Current Drug Targets – CNS & Neurological Disorders*, **2002**, *1*, 141.

11. Wadenburg, M-L G.; Kapur, S.; Soliman, A.; Jones, C.; Vaccarino, F. *Psychopharmacology*, **2000**, *150*, 422. Yokoi, F. *et. al.* *Neuropsychopharmacology*, **2002**, *27*, 248

12. (Risperidone) Kennis, L. E. J.; Vandenberk, J. US 4804663 (**1989**), EP 0196132 (**1992**).

13. Megens, A. A. H. P.; Kennis, L. E. J. *Prog. Med. Chem.* **1996**, *33*, 185.

14. *Drugs of the Future*, **1988**, *13*, 1052.

15. Hermecz, I.; Meszaros, Z. *Adv. Heterocyclic Chem.* **1983**, *33*, 241-330 (see p.255).

16. Willenbrock, H-J.; Wamhoff, H.; Korte, F. *Liebigs Ann. Chem.* **1973**, 103.

17. Fujıta, H.; Shimojı, Y.; Kojima, S.; Nıshino, H.; Kamoshıta, K.; Endo, K.; Kobayashı, S.; Kumakura, S.; Sato, Y. *Ann. Rep. Sankyo Res. Lab.* **1977**, *29*, 75.

18. Marquillas Olondrız, F.; Bosch Rovıra, A.; Dalmases Barjoan, P.; Caldero Ges, J. M. ES 2050069 (**1994**).

19. RPG Life Sciences Limited- Radhakrishnan, T. V.; Sathe, D. G.; Suryavanshi, C. V. WO 01/85731.

20. Teva- Krochmal, B.; Diller, D.; Dolitzky, B-Z.; Brainard, C. R. WO 02/14286.

21. (Olanzapine) Chakrabarti, J. K.; Hotten, T. M.; Tupper, D. E. US 5229382 (**1993**), EP 454436 (**1991**).

22. Chakrabarti, J. K.; Hotten, T. M.; Tupper, D. E. US 5627178 (**1997**).

23. Chakrabarti, J. K.; Horsman, L.; Hotten, T. M.; Pullar, I. A.; Tupper, D. E.; Wright, F. C. *J. Med. Chem.* **1980**, *23*, 878. (metabolites) Callıgaro, D. O.; Fairhurst, J.; Hotten, T. M.; Moore, N. A.; Tupper, D. E. *Bioorg. Med. Chem. Lett.* **1997**, *7*, 25.

24.   *Drugs of the Future*, **1994**, *19*, 114.

25.   Cen, J. *Chinese Journal of Pharmaceuticals*, **2001**, *32*, 391.

26.   Bunnell, C. A.; Hendriksen, B. A.; Larsen, S. D. US 5736541 (**1998**), EP 733635 (**1996**). (crystal form-II)

27.   Bunnell, C. A.; Larsen, S. D.; Nichols, J. R.; Reutzel, S. M.; Stephenson, G. A. EP 831098 (**1998**), US 6251895 (**2001**). (dihydrates B, D, E)

28.   Koprowski, R.; Reguri, B. R.; Chakka, R. WO 02/18390. (hydrates and Form-I)

29.   (Quetiapine) Warawa, E. J.; Migler, B. M. EP 240228 (**1987**), US 4879288 (**1989**).

30.   Barker, A. C.; Copeland, R. J. EP 282236 (**1988**).

31.   Warawa, E. J.; Migler, B. M. Ohnmacht, C. J.; Needles, A. L.; Gatos, G. C.; McLaren, F. M.; Nelson, C. L.; Kirkland, K. M. *J. Med. Chem.* **2001**, *44*, 372.

32.   Schmutz, J.; Kunzle, F.; Hunziker, F.; Burki, A. *Helv. Chim. Acta* **1965**, *48*, 336.

33.   *Drugs of the Future*, **1996**, *21*, 483.

34.   Bozsing, D.; *et. al.* WO 01/55125.

35.   (Ziprasidone) *Drugs of the Future*, **1994**, *19*, 560.

36.   Lowe, J. A.; Nagel, A. A. US 4831031 (**1989**).

37.   Howard, H. R.; Lowe, J. A.; Seeger, T. F.; Seymour, P. A.; Zorn, S. H.; Maloney, P. R.; Ewing, F. E.; Newman, M. E.; Schmidt, A. W.; Furman, J. S.; Robinson, G. L.; Jackson, E.; Johnson, C.; Morrone, J. *J. Med. Chem.* **1996**, *39*, 143.

38.   Yevich, J. P.; *et. al. J. Med. Chem.* **1986**, *29*, 359.

39.   Fox, D. E.; Lambert, J. F.; Sinay, T. G.; Walinsky, S. W. US 6111105 (**2000**). Walinsky, S. W.; Fox, D. E.; Lambert, J. F.; Sinay, T. G. *Org. Proc. Res. Dev.* **1999**, *3*, 126.

40.   Bowles, P. US 5206366 (**1993**). Allen, D. J. M.; Busch, F. R.; DiRoma, S. A.; Godek, D. M. US 5312925 (**1994**). (monohydrate)

41.   Urban, F. J. US 5359068 (**1994**). Urban, F. J.; Breitenbach, R.; Gonyaw, D. *Syn. Commun.* **1996**, *26*, 1629.

42.   Howard, H. R.; Shenk, K. D.; Smolarek, T. A.; Marx, M. H.; Windels, J. H.; Roth, R. W. *J. Labelled Compds. Radiopharm.* **1994**, *34*, 117.

43.   (Arıpıprazole) Oshiro, Y.: Sato, S.; Kurahashi, N. EP 0367141 (**1996**), US 5006528 (**1991**).

44.   *Drugs of the Future*, **1995**, *20*, 884.

45.   Oshiro, Y.: Sato, S.; Kurahashı, N.; Tanaka, T.; Kikuchi, T.; Tottori, K.; Uwahodo, Y.; Nishi, T. *J. Med. Chem.* **1998**, *41*, 658.

46.   Morita, S.; Kitano, K.; Matsubara, J.; Ohtani, T.; Kawano, Y.; Otsubo, K.; Uchida, M. *Tetrahedron*, **1998**, *54*, 4811.

Chapter 9. Atorvastatin Calcium (Lipitor®)

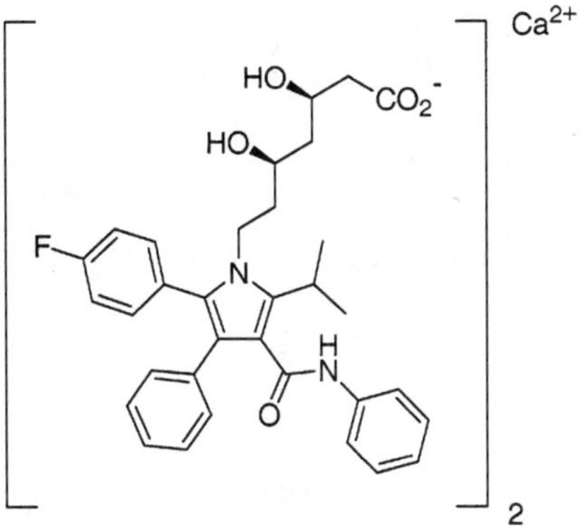

USAN: atorvastatin calcium
Trade Name: Lipitor ®
Company: Pfizer
Approval: 1996
M.W 1155.37

§9.1 Background[1,2]

Drugs that inhibit the enzyme HMG-CoA reductase (HMGR), the rate-limiting enzyme in cholesterol biosynthesis have become the standard of care for treatment of hypercholesterolemia due to the efficacy, safety and long-term benefits demonstrated in clinical trials employing these agents.[1] Currently, there are six compounds from this class that have been approved for marketing in the United States by the U.S. Food and Drug Administration. Three of these, lovastatin (2) marketed as MEVACOR®, simvastatin (3) marketed as ZOCOR®, pravastatin (4) marketed as PRAVACHOL®, are naturally occurring fungal metabolites or their semi-synthetic derivatives. The other three, fluvastatin (5) marketed as LESCOL®,

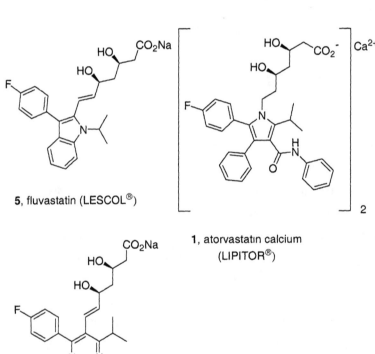

2, R=H, lovastatin (MEVACOR®)
3. R=CH₃,simvastatin (ZOCOR®)

4, pravastatin (PRAVACOL ®)

5, fluvastatin (LESCOL®)

1, atorvastatin calcium
(LIPITOR®)

6, rosuvastatin (CRESTOR®)

Figure 1. Marketed HMG-CoA reductase inhibitors.

atorvastatin calcium (1) marketed as LIPITOR® and rosuvastatin (6) marketed as CRESTOR®, are totally synthetic inhibitors. The fungal metabolites are complex hexahydronaphthalenes that are synthesized by fermentation. Fluvastatin, the first totally synthetic HMGR inhibitor was developed and is marketed as a racemic mixture.[2] Atorvastatin calcium was the first totally synthetic HMGR inhibitor to be developed and marketed as a single enantiomer, requiring considerable process development and physical infrastructure, especially in the area of low-temperature reactions.

§9.2 Synthesis of Racemic Atorvastatin[3–7]

Because the art of enantioselective drug synthesis was still in the early stages of development in the 1980s, especially on the scale required for a complex commercial process, many of the early synthetic routes produced racemic material. The first synthesis of atorvastatin was a small-scale synthesis by Parke-Davis discovery chemistry, utilizing a racemic synthesis followed by separation of diastereomers (Scheme 1).[3] Although several methods had been used previously to synthesize the central pyrrole ring of related, less highly substituted pyrrole HMGR inhibitors in discovery chemistry, synthesis of the penta-substituted pyrrole ring of atorvastatin was first accomplished by the 3 + 2 cycloaddition reaction of an acetylenic amide with an α-amido acid. Thus, alkylation of the ethylene glycol acetal of 3-amino-1-propanal (8) with α-bromoester 7 afforded α-amino acid ester 9. Reaction of compound 9 with isobutyryl chloride followed by hydrolysis of the ethyl ester provided the α-amido acid substrate 10 required for the 3 + 2 cycloaddition reaction. Heating 10 in acetic anhydride in the presence of excess 3-phenyl-propynoic acid phenylamide (11) afforded the 3 + 2 cycloaddition adduct 12, presumably through addition of the acetylene to the oxazolone intermediate derived from 10, followed by extrusion of CO_2.[4] As in similar 3 + 2 cycloadditions, the reaction was highly regioselective, forming essentially only one of the two possible regioisomeric products (12).[3,4] The masked aldehyde in 12 was most efficiently unveiled in a two-step, one-pot procedure by first converting 12 to the diethyl acetal, followed by acid hydrolysis to aldehyde 13. Reaction of 13 with the

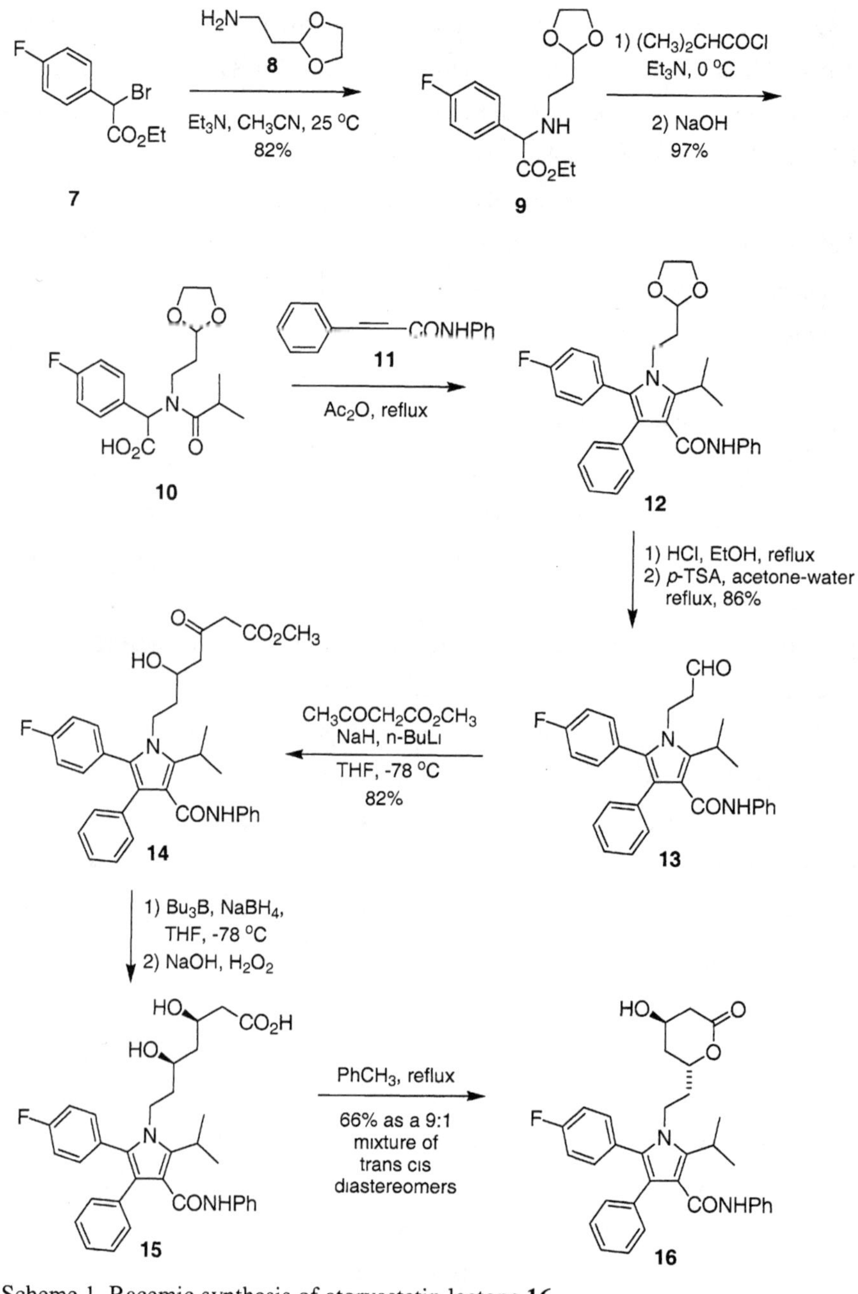

Scheme 1. Racemic synthesis of atorvastatin lactone **16**.

dianion of methyl acetoacetate under the conditions of Weiler[5] introduced all of the carbons required for the mevalonolactone as well as the 5-hydroxyl, although this center was introduced in a racemic fashion. Application of the diasteroselective reduction procedure of Narasaka and Pai[6] employing n-Bu$_3$B and NaBH$_4$ at low temperature (−78 °C) produced the 3,5-diol as a 9:1 mixture of the desired syn to the undesired anti-diol. Hydrolysis of the ester followed by lactonization in refluxing toluene produced the racemic lactone of atorvastatin as a 9:1 mixture of trans/cis diastereoisomers in an overall yield of 66% from **14**. One recrystallization of **16** raised this ratio to >97:3 trans/cis. Separation of the two enantiomers was accomplished by preparation of the diastereomeric (R)-α-methylbenzyl amides, separation by HPLC, hydrolysis and re-lactonization to produce 94% optically pure (+)-**16** (Scheme 2), a procedure used effectively to separate an analogous pair of HMGR inhibitors by Lynch, *et. al.*[7]

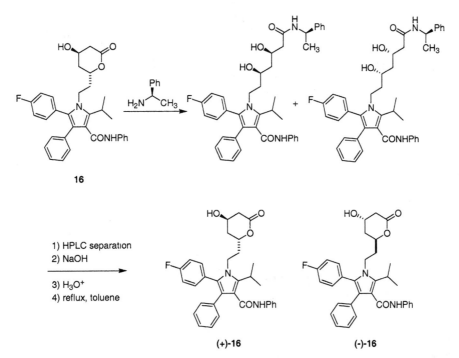

Scheme 2. Separation of the enantiomers of **16**.

§9.3 Enantioselective Syntheses of Atorvastatin Calcium[8-17]

Having identified the (+)-stereoisomer as the biologically active isomer, several independent enantioselective syntheses of this stereoisomer were developed. The initial synthesis developed in discovery chemistry employed the diastereoselective aldol condensation pioneered by Braun[8] as the key component. Thus, treatment of aldehyde **13** from the racemic synthesis with the magnesium enolate of (S)-(+)-2-acetoxy-1,1,2-triphenylethanol at −70 °C, afforded **17** in 60% yield as a 97:3 mixture of the R,S:S,S-diastereomers by HPLC (Scheme 3). Ester exchange employing sodium methoxide provided the methyl ester in quantitative yield. Reaction of this ester with three equivalents of lithio-t-butylacetate at −40 °C afforded the nearly enantiomerically pure t-butyl ester analog of racemic **14** in 75% yield.

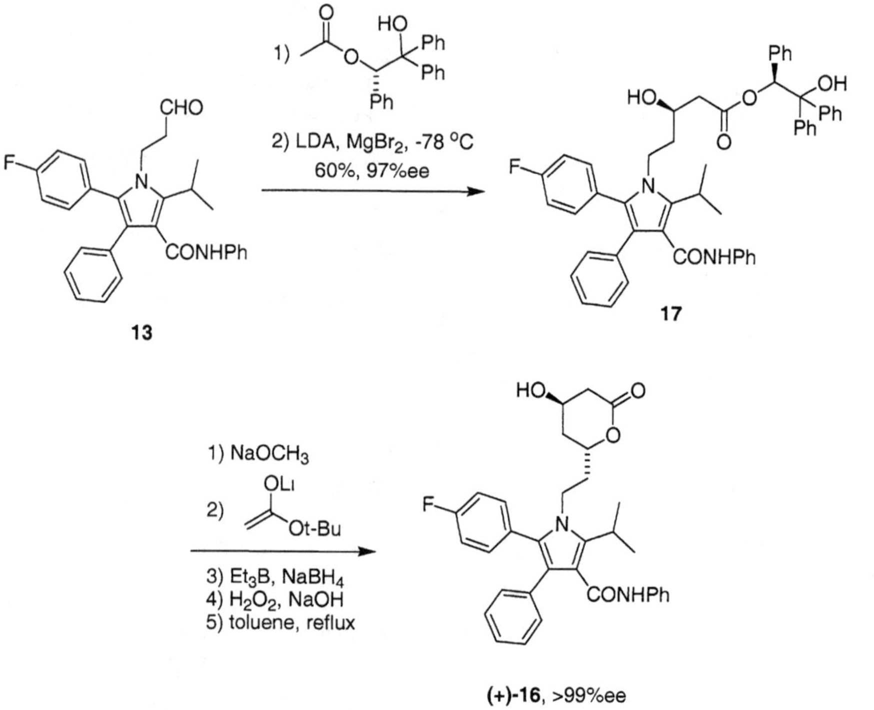

Scheme 3. Asymmetric synthesis of **(+)-16**.

Conversion of this intermediate to enantiomerically pure (+)-16 could be effected employing the same reaction sequence used to convert 14 to 16 in Scheme 1. Fortuitously, the *d,l*-pair crystallized out of ethyl acetate-hexanes and >99% enatiomerically pure (+)-16 could be isolated from the mother liquors as a foamy solid.

Although this route was successful in producing gram quantities of enantiomerically pure (+)-16, because of the linear nature of this route, the number of low-temperature reactions involved and the relatively low yields in some of the final steps (especially the final purification), its potential for scale-up to provide the kilogram quantities needed for further development was low. Thus, for the synthesis to be economically viable, an entirely different approach was taken by the Parke-Davis chemical development group of Butler, *et. al.*[9,10]

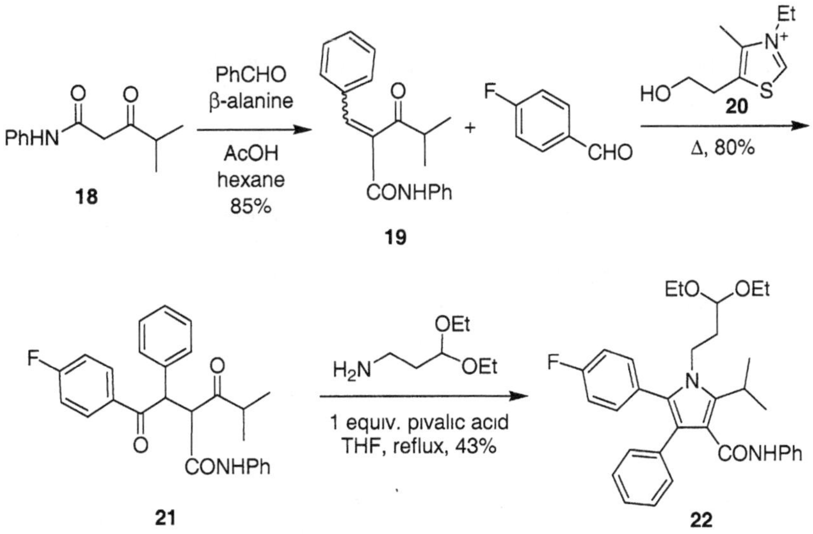

Scheme 4. Paal-Knorr synthesis of penta-substituted pyrrole 22.

A critical component of this effort was an extensive investigation of the classical Paal-Knorr pyrrole synthesis[11] that finally resulted in a successful cyclodehydration in a model system when a full equivalent of pivalic acid was used as catalyst (Scheme 4). Thus condensation of commercially available isobutyrylacetanilide 18 with benzaldehyde in the presence of β-alanine and acetic acid afforded the enone 19 in 85% yield.

Treatment of this enone with 4-fluorobenzaldehyde under the conditions of Stetter, utilizing *N*-ethylthiazolium catalyst **20** under anhydrous conditions produced the highly substituted 1,4-diketone **21** in 80% yield. Treatment of diketone **21** with the diethyl acetal of 3-amino-propanal under the carefully controlled conditions described above (1 equivalent of pivalic acid) afforded penta-substitued pyrrole **22** in 43% yield and, very significantly, demonstrated that a totally convergent synthesis was possible. With this result in hand, it became possible to envision a route in which a fully elaborated side-chain could be combined with the appropriate 1,4-diketone (*i.e.* **21**), to assemble the entire molecule in one operation.

To this end, several routes passing through the known (*S*)-methyl-4-bromo-3-hydroxybutyrate **26**,[12] an intermediate used in prior syntheses of HMGRIs, were developed. This key intermediate was derived most efficiently from isoascorbic acid as shown in Scheme 5.[12] Protection of **26** as the *t*-butyl-dimethylsilylether, followed by conversion to the nitrile provided an advanced intermediate (**27**) that could be taken in several directions.

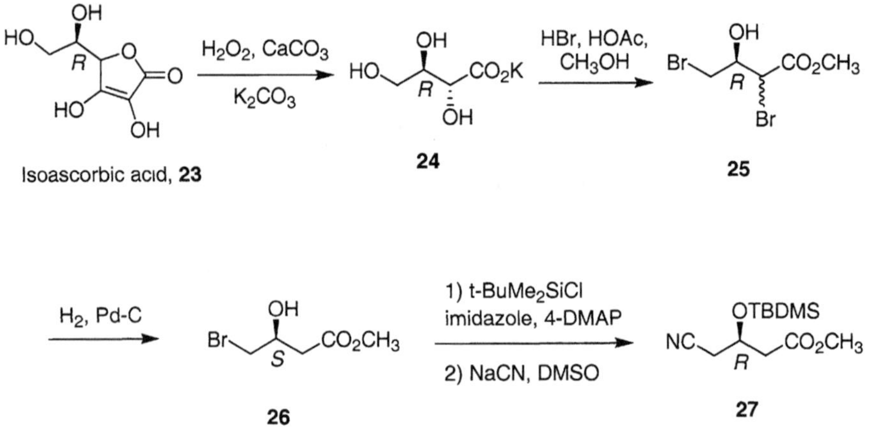

Scheme 5. Synthesis of (*S*)-methyl-4-bromo-3-hydroxybutyrate derivatives.

Thus, as shown in Scheme 6, sodium hydroxide mediated hydrolysis of **27** and chain extention by activation with *N,N*-carbonyldiimidazole followed by reaction with the magnesium salt of potassium *t*-butyl malonate and acidification followed by deprotection with buffered fluoride afforded the δ-hydroxy-β-ketoester **28**. Selective reduction of the ketone in **28** employing NaBH₄ and Et₂BOMe, a slight modification of the procedure used previously, gave the syn-1,3-diol.[13] Protection of the diol as the acetonide produced the nicely crystalline nitrile **29** in 65% yield and with diastereoselectivity in the range of 100:1. One recrystallization improved this ratio to > 350:1. Reduction of the nitrile with molybdenum-doped Raney-Nickel catalyst then afforded the desired side-chain (**30**) with outstanding enantiomeric excess (>99.5%) (Scheme 6).[9]

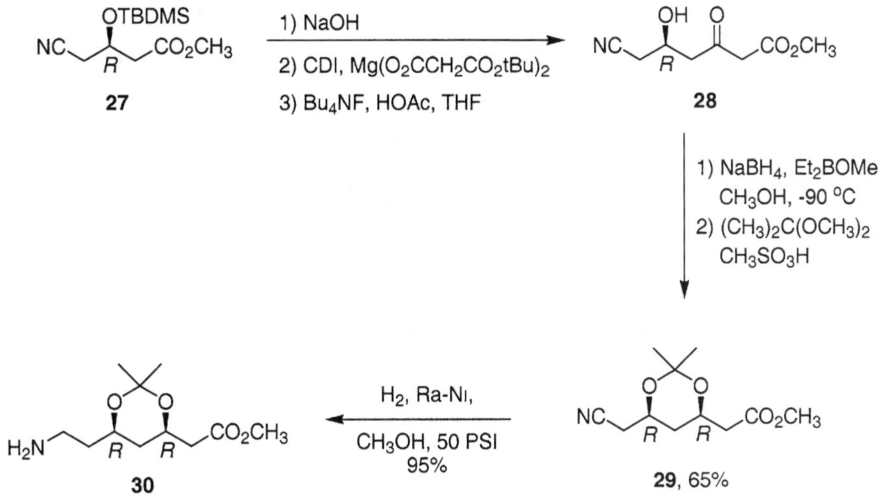

Scheme 6. Enantiospecific synthesis of side-chain **30**.

An alternate, shorter route was also developed which involved reaction of the alcohol derived from **27** with 3–4 equivalents of lithium *t*-butyl acetate to afford an excellent 75–80% yield of hydroxyketone **31** without the need for prior protection of the alcohol and with no detectable reaction with the nitrile (Scheme 7). Although these routes still involved a low-temperature reduction, both could be scaled to kilogram quantities.[9]

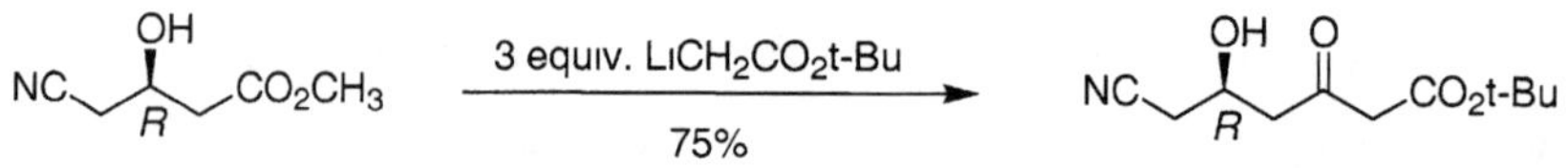

Scheme 7. Alternate side-chain synthesis.

A third route developed by this group started with the commercially available alcohol **32**,[14] a compound which has also been the subject of considerable process development due to its use as a common intermediate in the synthesis of several HMGR inhibitors.[15–17] Conversion of **32** to the 4-halo or 4-nitrobenzenesulfonate **33** followed by displacement with sodium cyanide provided **34** in 90% yield, which is the *t*-butyl-ester analog of **29**. It was noted that this procedure was most scaleable employing the 4-chlorobenzenesulfonate **33a** due to the instability of the 4-bromo and 4-nitro-analogs to aqueous hydrolysis. Ra-Ni reduction as before provided the fully elaborated side-chain **35** as the *t*-butyl ester (Scheme 8).

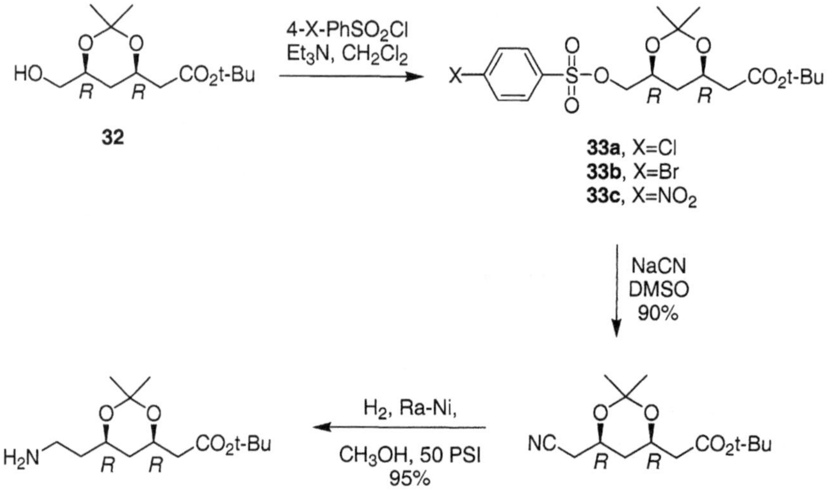

Scheme 8. Synthesis of side-chain **35** as the *t*-butyl ester.

With the fully functionalized, stereochemically pure side-chain **35** and the fully substituted diketone **21** in hand, these were reacted under very carefully defined conditions (1 equiv. pivalic acid, 1:4:1 toluene-heptane-THF) to afford a 75% yield of pyrrole **36** (Scheme 9). Deprotection and formation of the hemi-calcium salt produced stereochemically pure atorvastatin calcium in a convergent, high-yielding and commercially viable manner.[10]

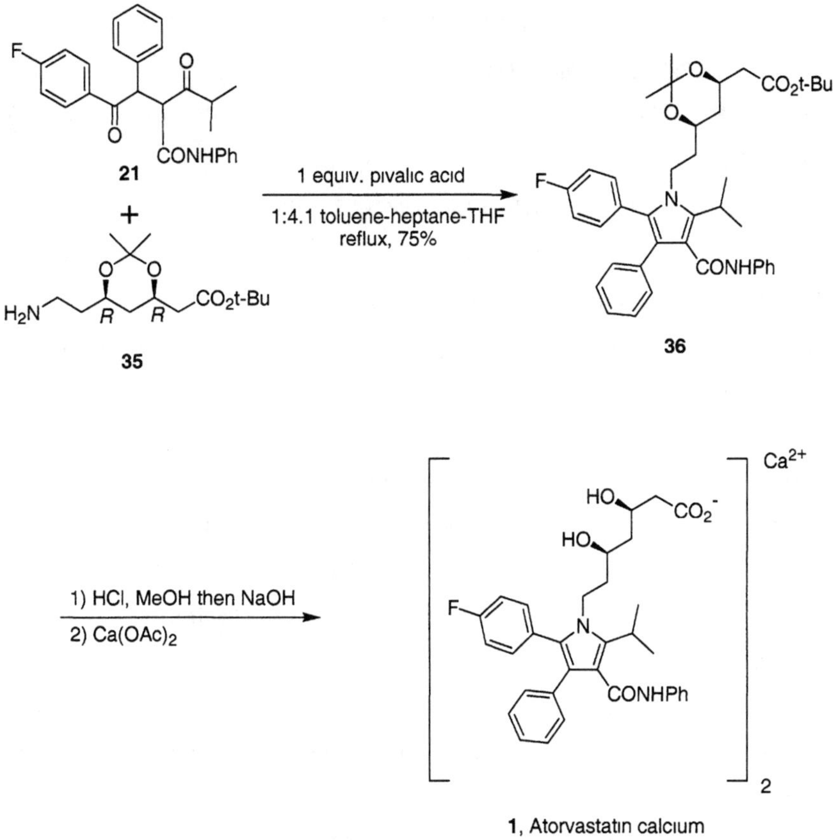

Scheme 9. Convergent, enantiospecific synthesis of atorvastatin calcium (**1**).

§9.4 References

1. Duriez, P. *Expert Opin. Pharmacother.* **2001**, 2, 1777–1794.

2. Repic, O.; Prasad, L.; Lee, G. T. *Organic Process Research & Development* **2001**, 5, 519–527.

3. Roth, B. D.; Blankley, C. J.; Chucholowski, A. W.; Ferguson, E.; Hoefle, M. L.; Ortwine, D. F.; Newton, R. S.; Sekerke, C. S.; Sliskovic, D. R.; Stratton, C. D.; Wilson, M. W. *J. Med. Chem.* **1991**, *34*, 357–366. Roth, B. D. US 4681893 (**1987**).

4. An extensive review of mesoionic heterocycles can be found in Newton, C. G.; Ramsden, C. A. *Tetrahedron* **1982**, *38*, 2965–3011.

5. Huckin, S. N.; Weiler, L. *J. Am. Chem. Soc.* **1981**, *103*, 6538–6539.

6. Narasaka, K.; Pai, H. C. *Chem. Lett.* **1980**, 1415–1418.

7. Lynch, J. E.; Volante, R. P.; Wattley, R. V.; Shinkai, I. *Tetrahedron Lett.* **1987**, 1385–1388.

8. Braun, M.; Devant, R. *Tetrahedron Lett.* **1984**, *25*, 5031–5034.

9. Browner, P. L.; Butler, D. E.; Deering, C. F.; Le, T. V.; Millar, A.; Nanninga, T. N.; Roth, B. D. *Tetrahedron Lett.* **1992**, *33*, 2279–2282.

10. Baumann, K. L.; Butler, D. E.; Deering, C. F.; Mennen, K. E.; Millar, A.; Nanninga, T. N.; Palmer, C. W.; Roth, B. D. *Tetrahedron Lett.* **1992**, *33*, 2283–2284. Roth B. D. US 5273995 (**1993**).

11. Knorr, L. *Ber.* **1885**, *18*, 299; Paal, C. *Ber.* **1885**, *18*, 367.

12. Sletzinger, M.; Verhoeven, T. R.; Volante, R. P.; McNamara, J. M.; Corley, E. G.; Lui, T. M. H. *Tetrahedron Lett.* **1985**, *26*, 2951–2954.

13. Chen, K. M.; Hardtmann, G. E.; Prasad, K.; Repic, O.; Shapiro, M. J. *Tetrahedron Lett.* **1987**, *28*, 155–158.

14. Available from Kaneka.

15. Takahashi, S., Yonetsu, K., Ueyama, N. European Patent Application 89123665.5, **1990**.

16. Wess, G.; Kesseler, K.; Baader, E.; Bartmann, W.; Beck, G.; Bergmann. A.; Jendralla, H.; Bock, K.; Holzstein, G.; Kleine, H.; Schnierer, M. *Tetrahedron Lett.* **1990**, *31*, 2545–2548.

17. Beck, G.; Jendralla, H.; Kesseler, K. *Synthesis* **1995**, 1014–1018.

Chapter 10. Antidepressants

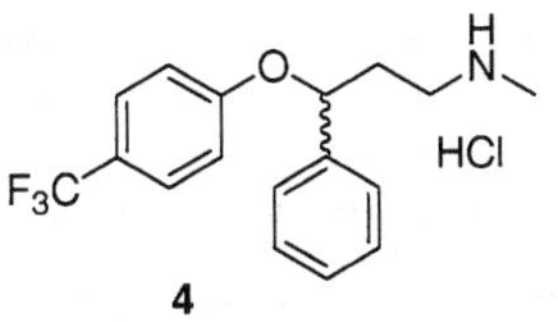

USAN: Fluoxetine Hydrochloride
Trade Name· Prozac®
Eli Lilly
Approval: 1987
M.W. 309 33

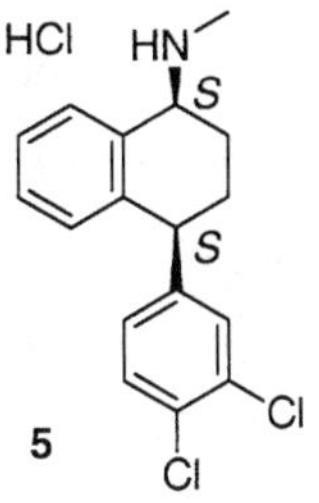

4

USAN: Sertraline Hydrochloride
Trade Name: Zoloft®
Pfizer
Approval: 1991
M.W. 306.23

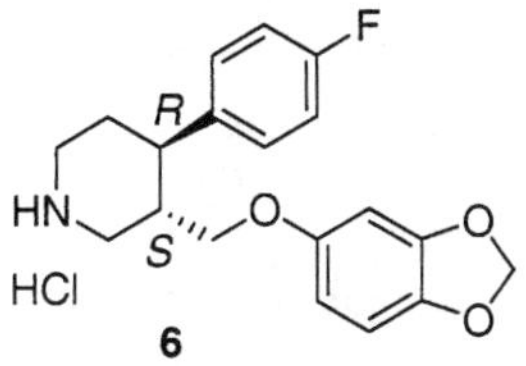

5

USAN: Paroxetine Hydrochloride
Trade Name: Paxil®
GlaxoSmithKline
Approval: 1992
M.W. 327.39

6

§10.1 Background

There are two major types of depression: major depressive disorder (MDD) and bipolar or manic-depressive illness. Both disorders are characterized by changes in mood as the primary clinical manifestation. Major depression is characterized by feelings of intense sadness and despair with little drive for socialization or communication. Physical changes such as insomnia, anorexia and sexual dysfunction can also occur.[1] Mania is characterized by excessive elation, irritability, insomnia, hyperactivity and impaired judgment. It may effect as much as 1% of the U.S. population.[2]

Major depressive disorder is among the most common psychiatric disorders in the United States, with an estimated 12-month prevalence of approximately 10% in the general population, and a prevalence of 12.9% and 7.7% in women and men, respectively. In terms of disease burden, as measured by Disability Adjusted Life Years (DALYS), MDD ranks as the fourth most costly illness in the world with estimated annual costs of depression in the United States amounting to approximately $43.7 billion.[3]

The exact causes of depression are not known. However, in the 1950s, it was observed that in addition to its other pharmacological properties, reserpine, a Rauwolfia alkaloid, induced a depressive state in normal patients and also depleted levels of the neurotransmitters, norepinephrine and serotonin. This single observation led to the hypothesis that the biological basis of major mood disorders may include abnormal monoamine neurotransmission.[4,5] Substances such as norepinephrine, serotonin and dopamine mediate neurotransmission. These substances are released from presynaptic neurons, cross the synaptic gap and interact with receptors on the postsynaptic cells. The synthesis, transmission and processing of these neurotransmitters provide a number of points of intervention through which a pharmacological agent may affect this transmission. Thus manipulation of neurotransmission has been the mainstay of antidepressant therapy for over half a century.

Several interventions are possible including 1) inhibiting enzymes that synthesize neurotransmitters, 2) preventing neurotransmitter storage in synaptic vesicles, 3) blocking the release of the neurotransmitter into the synaptic gap, 4) inhibiting neurotransmitter degradation, 5) blocking neurotransmitter reuptake, 6) agonism or antagonism of the postsynaptic receptor, 7) inhibiting signal transduction within the postsynaptic cell. Pharmacological agents have been identified that affect all of these processes, however, the mainstays of antidepressant therapy have been agents that affect neurotransmitter degradation and reuptake.

The first generation of antidepressants, MAO (monoamine oxidase) inhibitors, inhibited neurotransmitter degradation by inhibiting monoamine deoxidase, a flavin containing enzyme, found in the mitochondria of neurons and other cell types, that oxidatively deaminates naturally occurring sympathomimetic monoamines, such as norepinephrine, dopamine, and serotonin within the presynapse. In 1952, isoniazid and its isopropyl derivative, iproniazid (1), were developed for the treatment of tuberculosis, where it was subsequently found that these agents had a mood enhancing effect on tubercular patients. Elevation of mood is assumed to result from the accumulation of amines such as norepinephrine and serotonin in the central nervous system (CNS). It was later found that both were MAO inhibitors and their use in the treatment of depressed patients was a major milestone in modern psychiatry.[6] However, severe hepatoxicity was observed with iproniazid (**1**), and it was withdrawn from the U.S. market in 1961.

The tranquillizing properties of chlorpromazine (**2**), a conventional antipsychotic, have been known since the mid 1950s. Its discovery ignited a flurry of research into the search for close analogs that could be used in the treatment of a variety of psychiatric illnesses. A compound, imipramine (**3**), structurally related to chlorpromazine (**2**) and originally synthesized as an antihistamine, was subsequently shown to have pronounced antidepressant activity.[6] The introduction of this drug, Tofranil®, in 1958 ushered in the use of the tricyclic antidepressants (TCAs), whose mechanism of action is the inhibition of the reuptake of the biogenic amines.[7] When a neurotransmitter is released from a cell, it has only a short period of time to relay its signal before it is metabolized, via MAO, or is reabsorbed into the cell. All of the TCAs potentiate the actions of norepinephrine, serotonin and, to a lesser extent, dopamine. However, the potency and selectivity for inhibition of the uptake of norepinephrine, serotonin and dopamine vary greatly among the agents. Even though these agents are efficacious in their management of depression, they do have significant side-effects and toxicities (flushing, sweating, orthostatic hypotension, constipation) due to α-adrenergic blocking activity. All the TCAs are especially toxic in overdose, producing cardiac effects and seizures. These unwanted side-effects limit compliance, with as few as 1 in 17 patients completing a therapeutic-dosing regimen.[7]

Iproniazid, **1**   Chlorpromazine, **2**   Imipramine, **3**

The search for less toxic reuptake inhibitors led to the development of second-generation antidepressant agents known as the selective serotonin reuptake inhibitors (SSRIs). These agents differ from the older TCAs in that they selectively inhibit the reuptake of serotonin into the presynaptic nerve terminals, and therefore enhance synaptic concentrations of serotonin and facilitate serotonergic transmission. This increased neurotransmission and elevated synaptic levels of serotonin alleviate the symptoms and possibly the aetiology of depression. Relative to the TCAs they have a favorable side-effect profile and are much safer in overdose. However, they are generally not more efficacious than the TCAs; they exhibit a marked delay in onset of action and they have their own set of side-effects resulting from the non-selective stimulation of serotonergic receptor sites.[7-9]

The most widely known SSRIs are fluoxetine hydrochloride (4, Prozac®, Lilly), sertraline hydrochloride (5, Zoloft®, Pfizer) and paroxetine hydrochloride (6, Paxil®, GSK). Fluoxetine (4, Prozac®) was the first SSRI approved and effected a revolutionary change in the treatment of depression. In 2000, it was the most widely prescribed antidepressant drug in the United States with worldwide sales of $2.58B. Sertraline (5, Zoloft®) has been available in the United States since 1992 and had worldwide sales of $2.14B in 2000. Compared to fluoxetine (4), it has a shorter duration of action and fewer CNS activating side-effects such as nervousness and anxiety. Paroxetine (6, Paxil®) generated worldwide sales of $2.35B in 2000 and has a relatively benign side-effect profile, which favors its use with elderly patients.

§10.2 Synthesis of fluoxetine hydrochloride (4)

Fluoxetine (4) is marketed as a racemate. Both enantiomers display similar activity both in vitro (Ki of the (R)-(+)-enantiomer is 21 nM and the Ki of the (S)-(-)-enantiomer is 33 nM) and in vivo.[10] However, it is (S)-fluoxetine that is the predominant therapeutic enantiomer since it is eliminated more slowly that the (R)-enantiomer. However, recent debate has argued that the prolonged duration of action of the (S)-enantiomer contributes to the major side-effects of the drug.[11]

Molloy and Schmiegel described the original synthetic route to fluoxetine (4) (Scheme 1).[12] Mannich reaction of acetophenone (7) yielded the aminoketone 8.

Subsequent reduction of the prochiral ketone with diborane in THF and chlorination of the resulting secondary alcohol **9** provided a reactive benzylic chloride that underwent subsequent nucleophilic displacement with 4-trifluoromethylphenoxide to give **10**. Von Braun degradation of the *N,N*-dimethyl amine in **10**, via the *N*-cyano intermediate **11**, gave racemic fluoxetine (**4**).

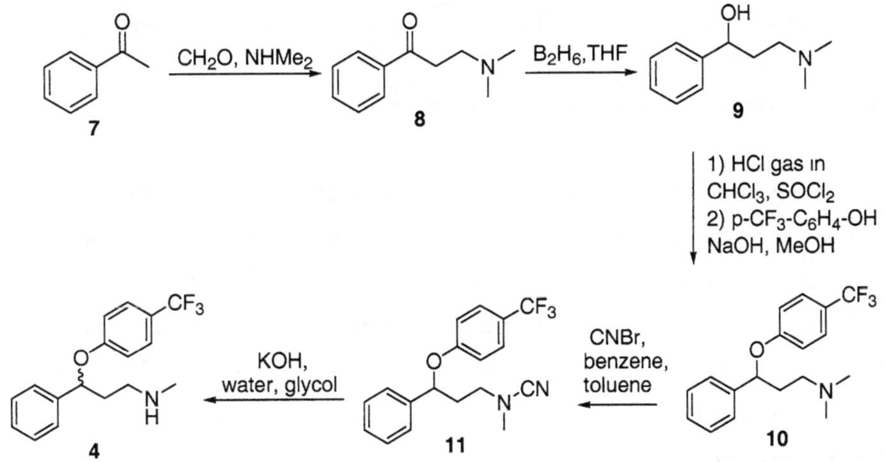

Scheme 1. The original synthesis of fluoxetine (**4**).

There has been much focused research on the enantioselective synthesis of both enantiomers of fluoxetine (**4**). Robertson *et al.* reported the first enantioselective synthesis of a fluoxetine enantiomer (Scheme 2).[10] In this synthesis, the asymmetry was introduced in the first step via an asymmetric reduction of 3-chloropropylphenyl ketone (**12**) using diisopinocamphenylchloroborane.[13] The (*S*)-alcohol **13** was obtained with an enantiomeric ratio of 96:4 (*S:R*). Treatment of **13** with NaI and aqueous methylamine introduced the amine functionality to give **14**. The alkoxide of **14** was added to 4-fluorobenzotrifluoride to give (*S*)-fluoxetine ((*S*)-**4**) with an enantiomeric ratio of 96:4 (*S:R*), thus indicating that no racemization had occurred during the synthesis. The principal drawback of this method is that it could only be used to provide the (*S*)-enantiomer. The authors utilized classical resolution methods employing *D*-and *L*-mandelic acid salts to give the (*R*)-enantiomer.[10,14]

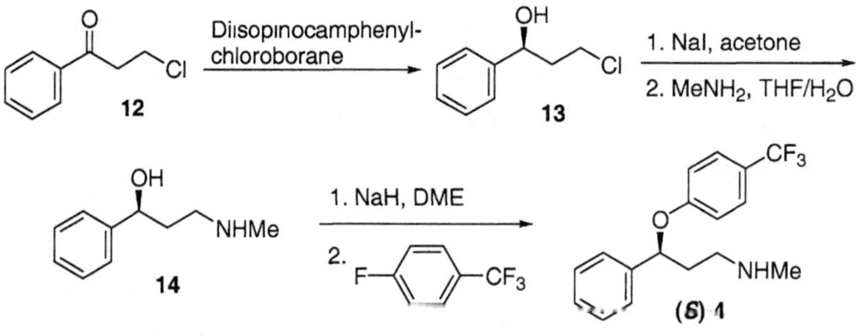

Scheme 2. Asymmetric synthesis of (S)-fluoxetine ((S)-4).

Corey and Reichard described a more efficient synthesis of the fluoxetine enantiomers.[15] This synthesis features a catalytic reduction of prochiral ketone 12 to install the correct absolute stereochemistry at C-3. In this respect this method is very similar to the one previously described by Robertson et al.[10] However, the major advantage of the Corey procedure is that the reduction utilizes chiral enzyme-like catalysts to induce the correct stereochemistry and both enantiomers of the catalyst are available (15 and 16).

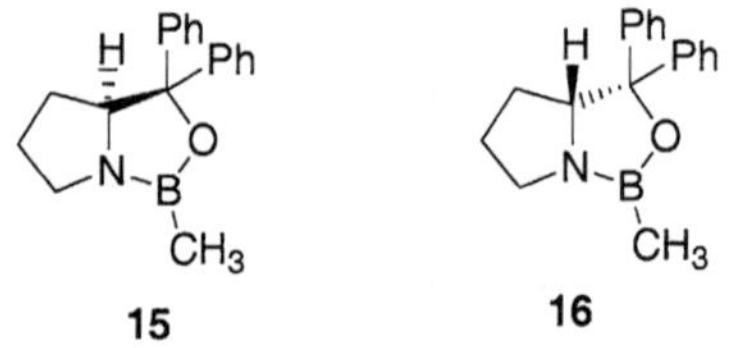

In a closely related reaction, asymmetric reduction of 3-(N-benzyl-N-methylamino)propiophenone using H$_2$, [Rh(COD)Cl]$_2$ and the chiral phosphine (2S,4S-(1-(N-methyl-carbamoyl)-4-(dicyclohexylphosphino)-2-[diphenylphosphino)methyl]-pyrrolidine ((2S,4S)-MCCPM), followed by debenzylation and addition of the alcohol to 1-chloro-4-(trifluoromethyl)benzene gave (R)-fluoxetine (4-(R)).[16]

Many other methods utilizing a three-carbon-chain segment have been employed in the syntheses of the fluoxetine enantiomers. Chirality has been established by enzymatic reduction,[17,18] lipase mediated enzymatic resolution,[19-22] oxidative kinetic resolution,[23] Sharpless asymmetric hydroxylation,[24] asymmetric carbonyl-ene reaction[25] and a stereospecific ruthenium-catalyzed allylic alkylation.[26]

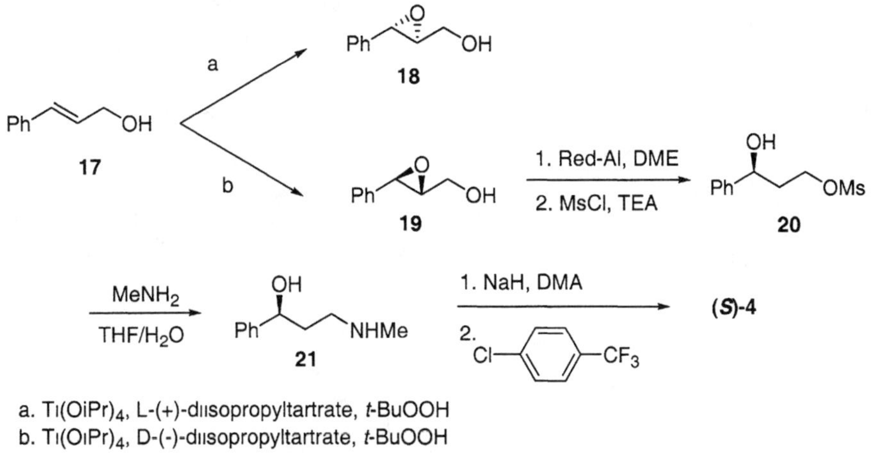

a. Ti(OiPr)$_4$, L-(+)-diisopropyltartrate, t-BuOOH
b. Ti(OiPr)$_4$, D-(-)-diisopropyltartrate, t-BuOOH

Scheme 3. Sharpless asymmetric epoxidation route to the fluoxetine enantiomers.

Work continues on the discovery of a general route to the enantiomers that should be adaptable to the efficient preparation of each of the fluoxetine enantiomers. The first such synthesis was described by Sharpless and Gao (Scheme 3).[27] This method utilized the asymmetric epoxidation of cinnamyl alcohol and subsequent regioselective reduction of the resulting chiral epoxide with 'Red-Al'. Asymmetric epoxidation of cinnamyl alcohol (17) yielded either chiral epoxy alcohol 18 or 19 dependent upon the specific choice of epoxidation catalyst. (S)-Fluoxetine required the use of D-(−)-diisopropyl tartrate, while the (R)-enantiomer required L-(+)-diisopropyl tartrate. Chiral epoxide 19 was regioselectively reduced with 'Red-Al' to give the 1,3-diol. Selective mesylation of the primary alcohol with one equivalent of mesyl chloride provided the sulfonate 20, which was then displaced with methylamine in aqueous THF to give 21. Subsequent generation of the alkoxide followed by treatment with p-chlorobenzotrifluoride gave (S)-fluoxetine ((S)-4) in an overall yield of 49%. The (R)-(+)-enantiomer ((R)-4) was prepared in a similar manner from chiral epoxide 18.

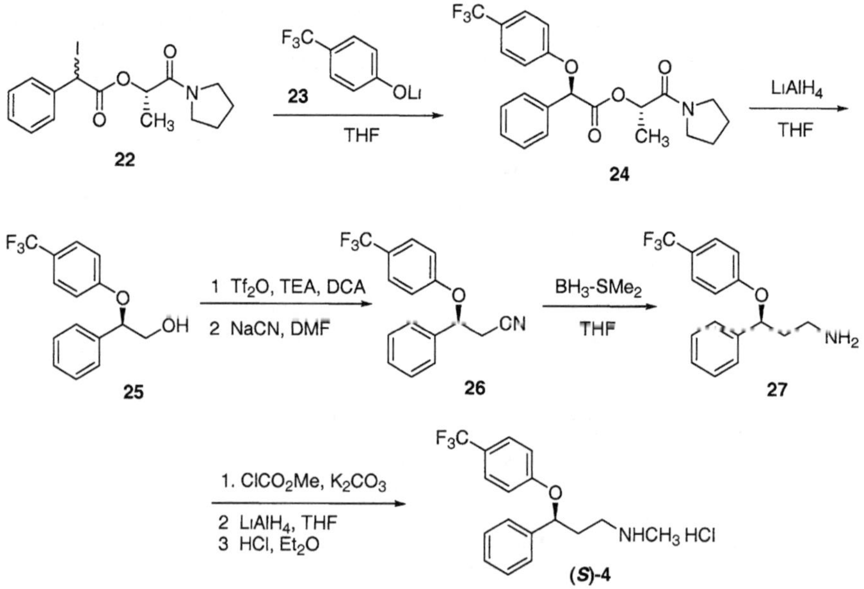

Scheme 4. Synthesis of (*S*)-fluoxetine from iodoester **22**.

Interestingly, all of the aforementioned syntheses involve the formation of a 3-hydroxy-3-phenyl substituted propylamine or closely related derivative. Devine *et al.* have described a highly stereoselective coupling reaction between racemic α-haloacids and aryloxides mediated by a pyrrolidine derived (*S*)-lactamide auxiliary (Scheme 4).[28] Iodoester **22** was readily synthesized from the bromoacid and the chiral auxiliary. This then underwent a highly stereoselective reaction with the lithium alkoxide **23** to give the coupled product **24** as a 98:2 mixture of diastereomers, which was further purified by recrystallization of the corresponding hydrochloride salt. Reduction with LiAlH₄ in THF gave the alcohol **25**, which was converted to the nitrile **26** via the corresponding triflate. Nitrile **26** was readily reduced with borane in refluxing THF to the primary amine **27**. *N*-methylation was accomplished by first forming the carbamate under Schotten-Baumann conditions. Reduction of the carbamate followed by acidification gave (*S*)-(-)-fluoxetine ((**S**)-**4**) in 37% overall yield.

§10.3 Synthesis of sertraline hydrochloride (5)

Early studies from the Pfizer laboratories had revealed that compounds from a series of *trans*-1-amino-4-phenyl-tetralins possessed potent norepinephrine (NE) uptake blocking activity. The activity was highly specific for the (1*R*, 4*S*)-enantiomer and was confined to the *trans* derivatives. The corresponding (1*S*, 4*R*)-enantiomer was much less active and the diastereomeric *cis* racemates were inactive at blocking NE uptake. It was subsequently shown that many compounds from the diastereomeric cis series were unexpectedly potent and selective inhibitors of serotonin (5-HT) uptake, thus differentiating these compounds from the *trans* compounds. One of these compounds, sertraline (5), was originally discovered as a racemic mixture. Resolution showed that the (+)-enantiomer was several times more selective for 5-HT uptake blocking activity than the (-)-isomer. The (+)-enantiomer was subsequently shown to possess the in vivo behavioral effects expected of a potent and selective 5-HT blocker. Thus, as opposed to fluoxetine (4), sertraline (5) is a single enantiomer with the *cis*-(1*S*, 4*S*) absolute configuration.[29-31]

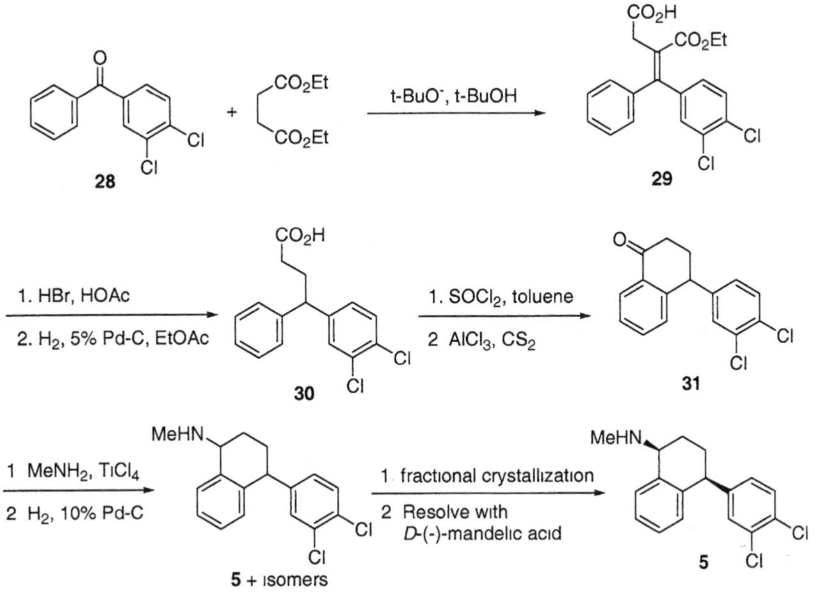

Scheme 5. Original Pfizer synthesis of sertraline (5).

The original synthesis of sertraline utilized the Stobbe reaction to couple the benzophenone **28** and diethylsuccinate to yield the mono acid **29** (Scheme 5). This was then hydrolyzed and decarboxylated under strongly acidic conditions to yield the but-3-enoic acid which was reduced with hydrogen over a palladium catalyst to give the 4,4-diarylbutanoic acid **30**. Treatment with thionyl chloride gave the acid chloride, which cyclized, under Friedel-Crafts conditions, onto the more reactive aryl ring to yield the tetralone **31**. Condensation with methylamine in the presence of titanium tetrachloride, followed by catalytic reduction of the resulting imine gave a 70:30 mixture of *cis* and *trans* amines. The *cis* form was purified as its hydrochloride salt by fractional crystallization and it was subsequently resolved with D-(-)-mandelic acid to give (+)-(1S,4S)-sertraline (**5**).[32]

Tetralone **31** could also be synthesized much more efficiently by employing a chemoselective ketone reduction of **32** to give the lactone **33**. A double Friedel-Crafts alkylation/acylation sequence employing a variety of Lewis or protic acids and benzene gave the tetralone **31** directly. Triflic acid and HF produced the highest yields of tetralone, presumably through the intermediacy of the diaryl acid **34** (Scheme 6).[29,33]

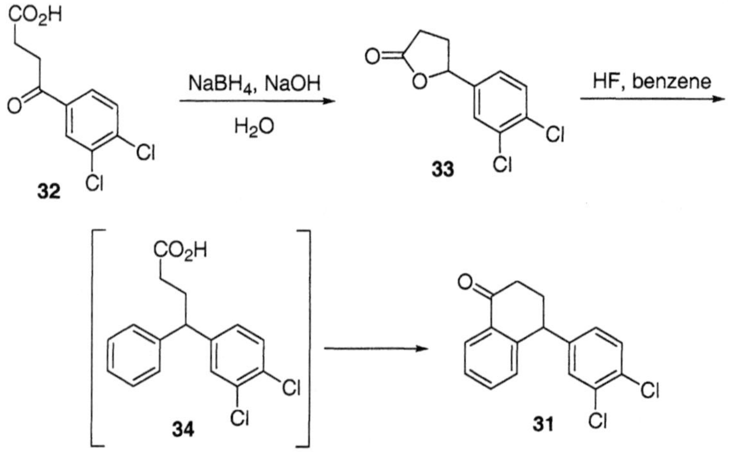

Scheme 6. Synthesis of tetralone **31** via a double Friedel-Crafts reaction sequence.

Qualich and Woodall described the first asymmetric synthesis utilizing a catalytic enantioselective reduction of the ketoester **35** with (*S*)-terahydro-1-methyl-3,3-diphenyl-1*H*,3*H*-pyrrolo[1,2-c][1,3,2]oxazaborole (CBS) to give the desired hydroxyester **36** (90% ee). After mesylation, S_N2 displacement with a higher-order cuprate derived from copper cyanide gave the diaryl *t*-butyl ester **37** with good chirality transfer. Intramolecular Friedel-Crafts cyclization gave the tetralone **31** in 90% ee (Scheme 7).[34]

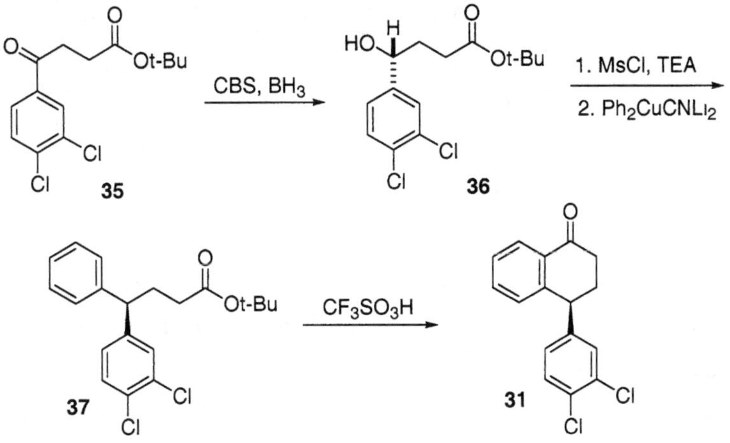

Scheme 7. Enantioselective synthesis of tetralone **31**.

The intramolecular C-H insertion reaction of phenyldiazoacetates on cyclohexadiene, utilizing the catalyst $Rh_2(S\text{-}DOSP)_4$, leads to the asymmetric synthesis of diarylacetates (Scheme 8). Utilizing the phenyldiazoacetate **38** and cyclohexadiene, the C-H insertion product **39** was produced in 59% yield and 99% ee. Oxidative aromatization of **39** with DDQ followed by catalytic hydrogenation gave the diarylester **40** in 96% ee. Ester hydrolysis followed by intramolecular Friedel-Crafts gave the tetralone **31** (96% ee) and represents a formal synthesis of sertraline (**5**).[35] Later studies utilized the catalyst on a pyridine functionalized highly cross-linked polystyrene resin.[36]

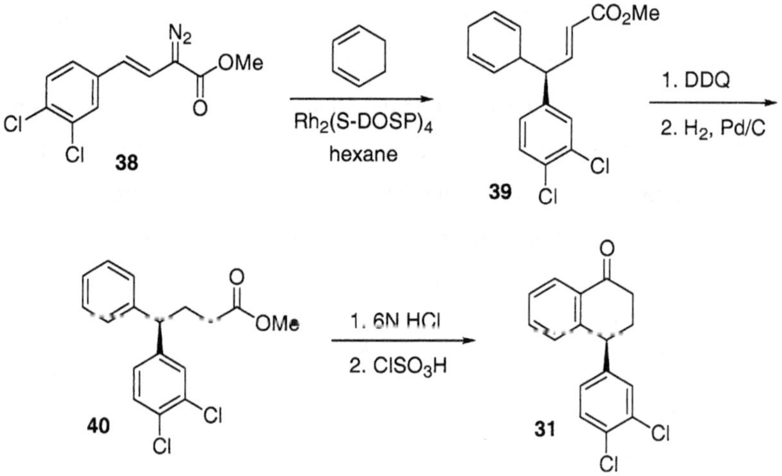

Scheme 8. Enantioselective C-H insertion route to tetralone **31**.

In another catalytic asymmetric synthesis, addition of the carbene generated from the diazoester **41** in the presence of the rhodium catalyst **42** to styrene gave the cyclopropane **43** in 94% ee. Oxidative degradation to the malonic acid monoester and methylation gave the dimethyl ester **44**. Homoconjugative addition of the cuprate **45** (Ar = 3,4-dichlorophenyl) gave the malonic ester **46**, which was subsequently hydrolyzed and decarboxylated to give the 4,4-diarylbutyric acid **40** (see Scheme 8). Cyclization of this acid occurred smoothly upon reaction with chlorosulfonic acid to give the tetralone **31** in 100% ee (Scheme 9).[37]

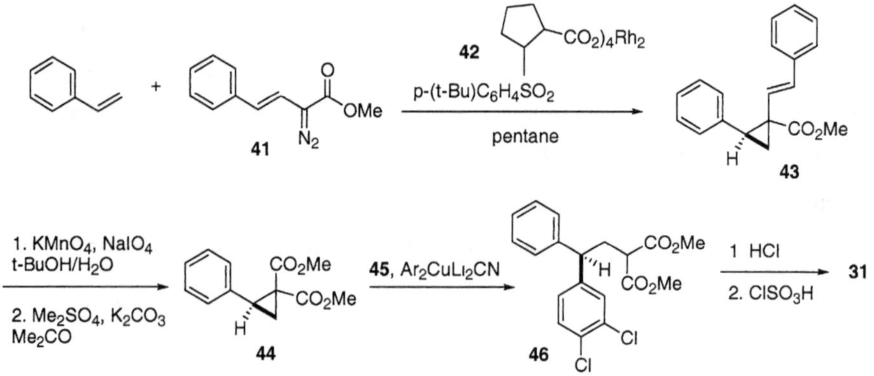

Scheme 9. Enantioselective cyclopropanation route to tetralone **31**.

Lautens and Rovis developed a general strategy to the enantioselective synthesis of the tetrahydronaphthalene core found in sertraline and a number of other bioactive molecules. They found that oxabenzonorbornenes undergo highly enantioselective reductive ring opening in the presence of Ni(COD)₂/(S)-BINAP and DIBAL-H. Thus oxabenzonorbornene (**47**) underwent this nickel catalyzed hydroalumination reaction to give the dihydronaphthalenol **48** in 91% ee. Protection of the alcohol and bromination provided a dibromide, that was not isolated, but was subsequently treated with DBU to give the vinyl bromide **49**. This was then subjected to a palladium catalyzed Stille cross-coupling with the requisite arylstannane to yield **50**. Later studies by the same authors utilized a Suzuki coupling with the requisite arylboronic acid to yield **50**. After deprotection, a selective reduction of alcohol **51** using hydrogen gas and Crabtree's catalyst ([Ir(COD)pyPCy₃]PF₆) gave alcohol **52** as a 28:1 mixture of diastereomers. Oxidation to the ketone **31** using manganese dioxide represented the completion of the formal synthesis of sertraline (**5**) (Scheme 10).[38,39]

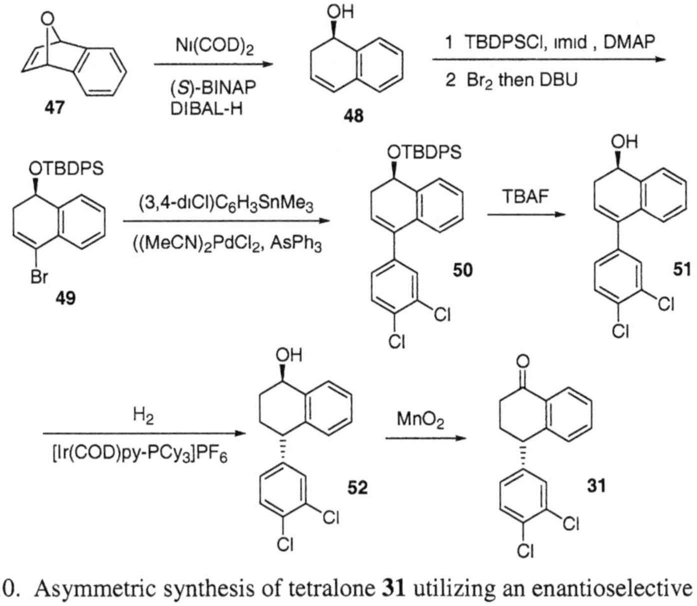

Scheme 10. Asymmetric synthesis of tetralone **31** utilizing an enantioselective Ni-catalyzed ring-opening of oxabenzonorbornenes.

§10.4 Synthesis of paroxetine hydrochloride (6)

Paroxetine hydrochloride (6) is an enantiomerically pure (-)-*trans*-3,4-disubstituted piperidine. The first synthesis of paroxetine (6) utilized an intermediate that allowed the ready separation of the *cis*- and *trans*-isomers each in racemic form (Scheme 11). Treatment of arecoline (53) with 4-fluorophenyl magnesium bromide gave a mixture of the *trans*- and *cis*-aryl piperidines 54 and 55. The *cis*-isomer 55 could be isolated pure by equilibration with NaOMe. The *trans*-isomer was available only by separation of the original mixture. The *cis*-isomer 55 was hydrolyzed with aqueous HCl and the resulting acid was treated with thionyl chloride to give 56. Acid chloride 56 was reacted with (-)-menthol to give the menthol ester 57 and the corresponding diastereomer, which was separated by fractional distillation. The (-)-*cis*-isomer was hydrolyzed under acidic conditions and reduced to the carbinol 58. Chlorination and treatment with sodium 3,4-methylenedioxyphenoxide (sesamol) gave 59. Demethylation could be achieved under mild conditions by treatment with vinylchloroformate to yield the vinylurethane 60. This was hydrochlorinated with HCl gas to yield the corresponding chloroethylurethane, which was subsequently hydrolyzed in refluxing methanol to yield *cis*-6 (Scheme 11).[40] Engelstoft and Hansen have developed a much more efficient demethylation method using cyanogen bromide and LiAlH$_4$.[41] The *trans*-isomer 54 could be converted to paroxetine (6) in a similar fashion.

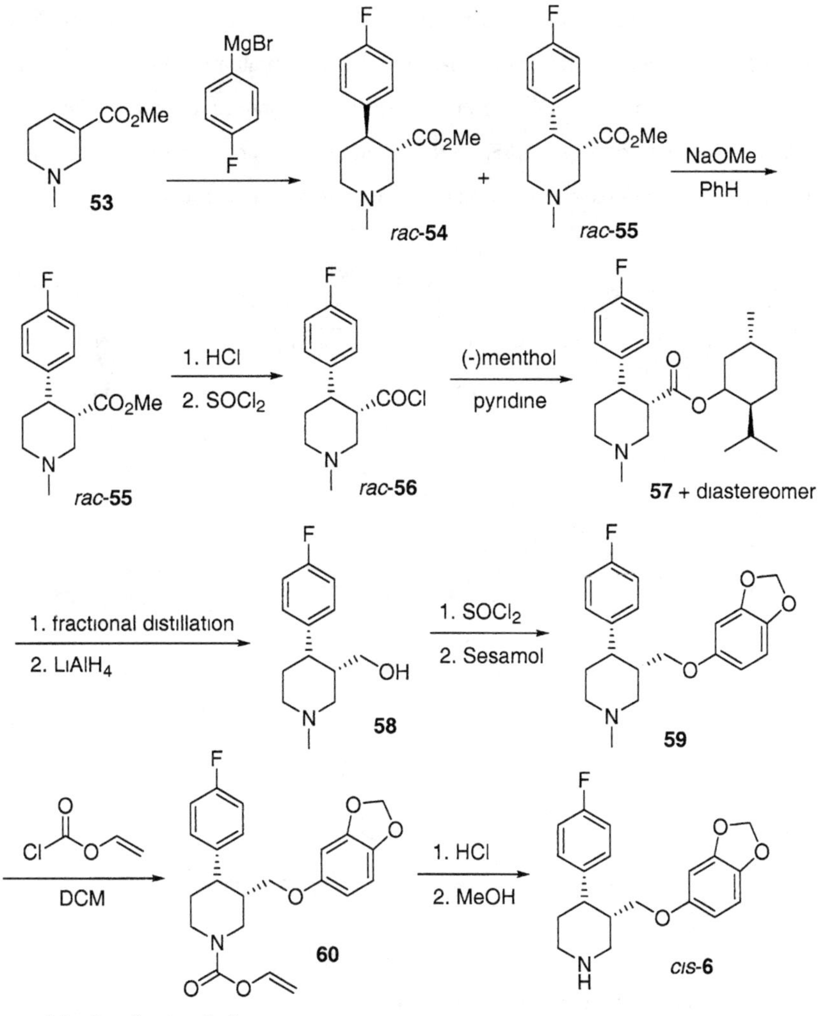

Scheme 11. Synthesis of *cis*-paroxetine.

In a closely related asymmetric reaction, the required absolute stereochemistry at C-4 was established via a Michael addition of a cuprate reagent to a dihydropiperidinone (Scheme 12).[42] The stereochemistry at C-3 was introduced in the form of piperidinone **61**, a compound readily available from (S)-glutamic acid. Protection of both the amino and alcohol functionalities was achieved using standard reaction conditions to give **62**. Introduction of the Δ^3-double bond was accomplished via phenylselenation of the lithium enolate of **62**, followed by treatment with hydrogen peroxide to yield the Michael acceptor **63**. Conjugate addition of the 4-fluorophenyl cuprate reagent in the presence of TMSCl gave **64** in >96% de. Reduction of the ketone functionality followed by deprotection of the alcohol gave **65**. SN₂ coupling between alcohol **65** and piperonyl chloride in basic DMSO gave the *N*-Boc protected paroxetine analog **66**.

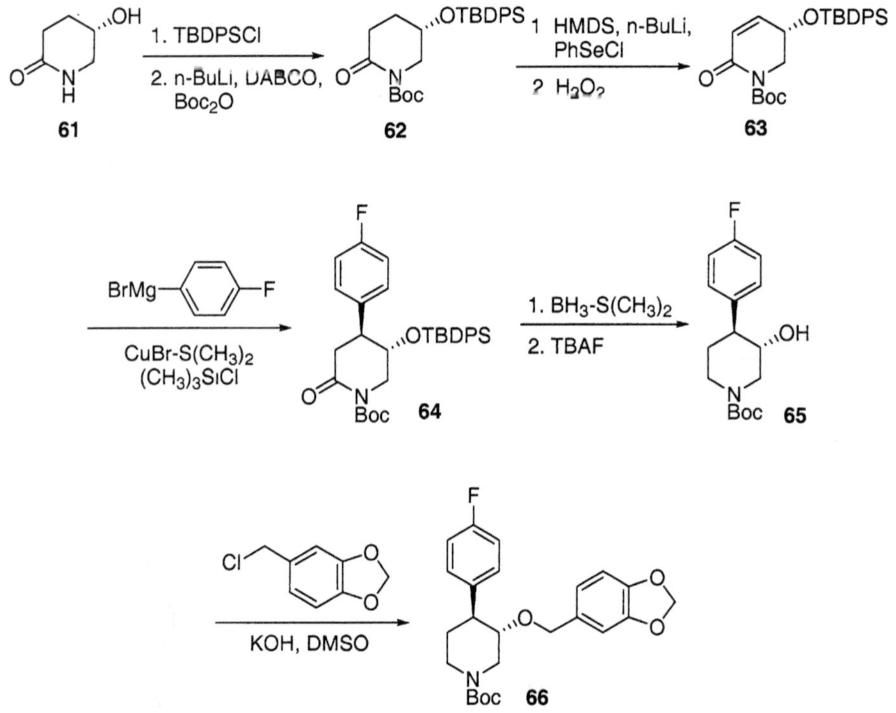

Scheme 12. Synthesis of paroxetine analog **66**.

Amat and co-workers have described an enantiodivergent synthesis of both (+)- and (-)-paroxetine (Scheme 13).[43] Thus, treatment of (*R*)-phenylglycinol and methyl 5-oxopentanaote gave an 85:15 mixture of bicyclic lactams *cis*-**67** and *trans*-**68**, which can be separated by chromatography or by equilibration to give the *trans*-isomer **68**. The requisite α,β-unsaturated lactam **70** was produced by sequential treatment of *trans*-**68** with LHMDS, methylchloroformate and phenylselenium bromide followed by ozonolysis of the resulting selenide **69**. Conjugate addition of the requisite arylcyanocuprate gave **71** and **72** in a 97:3 ratio of diastereomers. Alane reduction of **71** cleaved the oxazolidine ring and reduced the ester and lactam carbonyl groups to give the enantiopure *trans*-piperidine **73**. Hydrogenolysis provided alcohol **74**, which was mesylated and treated with the sodium salt of sesamol to afford **75**, which on treatment with TFA gave (+)-paroxetine (**6**).

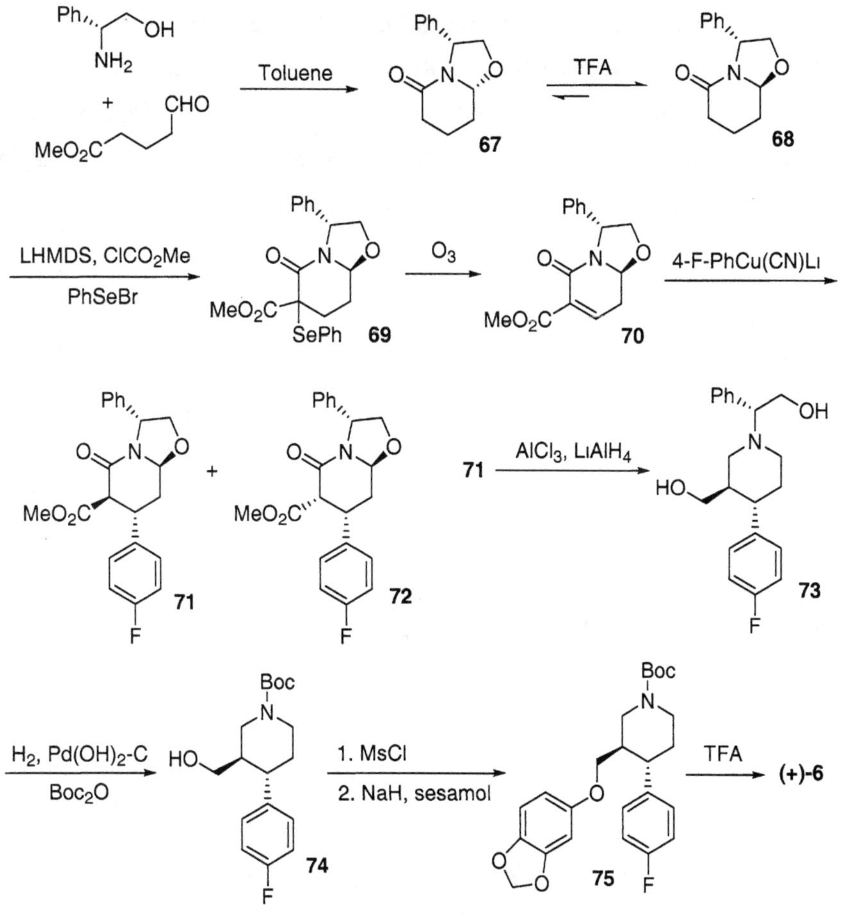

Scheme 13. Asymmetric synthesis of (+)-paroxetine (**6**).

There have been a number of syntheses where the requisite stereochemistry has been introduced by the asymmetric desymmetrization of a variety of glutaric acid analogs

(Scheme 14).[44] The bis ester **76** was produced by reaction of *p*-fluorobenzaldehyde with ethyl acetoacetate and NaOH followed by esterification. Pig liver esterase (PLE) mediated enzymatic hydrolysis gave the monomethyl glutaric acid **77** in 86% yield with an *ee* of 95%. Deprotonation of the acid followed by reduction with LiBH$_4$ gave a boronate intermediate, which was alkylated and then quenched with methanol to give the δ-hydroxy ester **78**. Alcohol **78** was mesylated and treated with benzylamine to give the aminoester, which cyclized to give the lactam **79**. Acylation gave **80**, which was reduced with LiAlH$_4$ to give the key aminoalcohol **81**. Etherification with sesamol gave **82** and subsequent hydrogenolysis of the benzyl group gave (−)-paroxetine (**6**) (Scheme 14).[44] Interestingly, closely related aminoalcohols have been resolved by lipase-catalyzed esterification using commercially available cyclic anhydrides as acyl donors.[45]

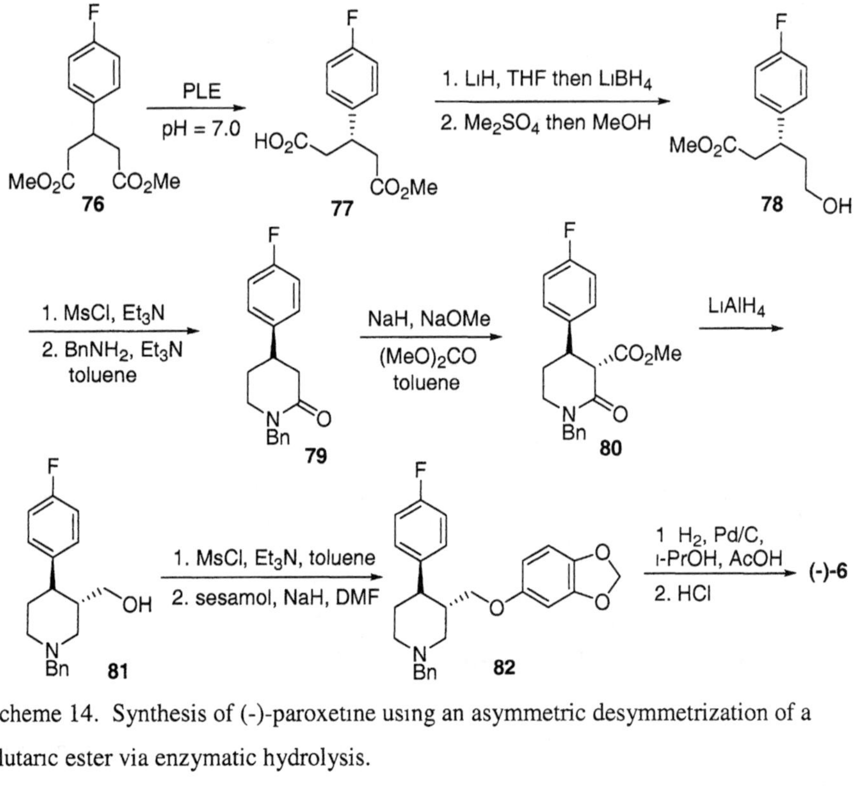

Scheme 14. Synthesis of (-)-paroxetine using an asymmetric desymmetrization of a glutaric ester via enzymatic hydrolysis.

A similar sequence was reported where the asymmetry was introduced by the reaction of *meso*-3-substituted glutaric anhydrides and (*S*)-methylbenzylamines to give diastereomeric hemiamides that could be separated by recrystallization.[46] The asymmetric desymmetrization of certain 4-aryl substituted glutarimides has also been accomplished with high levels of selectivity (up to 97% ee) by enolization with a chiral bis-lithium amide base. The selectivity of the reaction was shown to be the result of asymmetric enolization, followed by a kinetic resolution.[47]

Several other methods have been utilized to produce the aminoalcohol **81**. A ring expansion of prolinol **84** (derived from pyroglutamic acid (**83**) in eight steps) gave the tri-substituted chloropiperidine **85**. Dechlorination of **85** was accomplished under reductive radical conditions to give the aminoester **86**, which was reduced to the aminoalcohol **81**, thus completing the formal synthesis of (-)-paroxetine (**6**) (Scheme 15).[48]

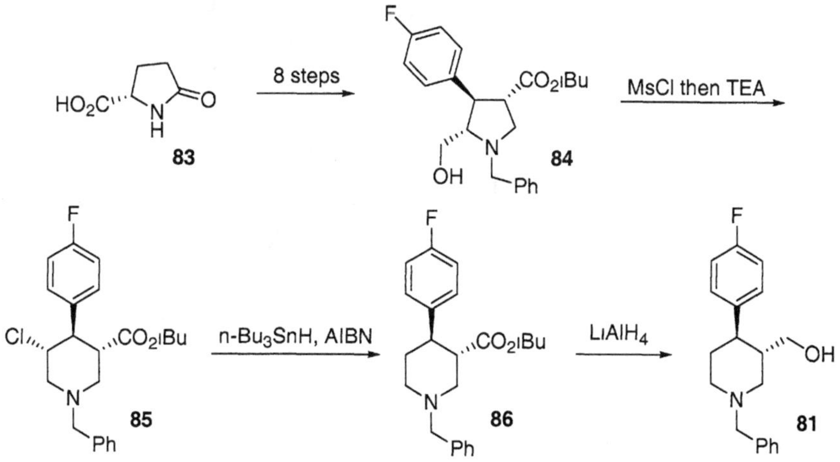

Scheme 15. Formal synthesis of (-)-paroxetine via piperidine **81**.

Beak and co-workers have also produced the key alcohol intermediate **74** by the sparteine-mediated lithiation and conjugate addition of allylamines to nitroalkenes to give Z-enecarbamates in good yields with high enantio- and diastereoselectivity (Scheme 16).[49] Thus treatment of the allylamine **87** with n-BuLi in the presence of (−)-sparteine followed by conjugate addition to nitroalkene **88** gave the desired enecarbamate **89** in

83% yield as a single diastereomer. Hydrolysis and reduction of the corresponding aldehyde gave the nitro alcohol **90** in 88% yield. Reduction of the nitro functionality by transfer hydrogenation and subsequent Boc-protection gave **91** in 95% yield. Cyclization and deprotection gave alcohol **74** in 83% yield, which could be carried on to give (-)-paroxetine (**6**).

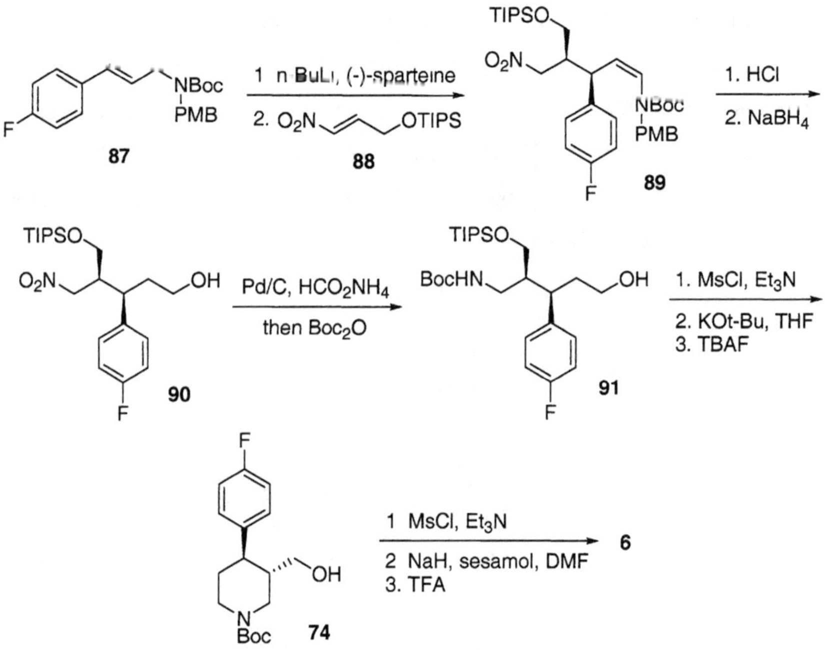

Scheme 16. Synthesis of (-)-paroxetine utilizing a sparteine-mediated lithiation and conjugate addition reaction.

§10.5 References

1.   Baldessarini, R. J. in *Goodman and Gilman's The Pharmacological Basis of Therapeutics* (**1985**) (Gilman, A.G., Goodman, L.S., Rall, T.W., Murad, F (Eds)) 7th Edition, pp 412–432, Macmillan, New York.

2.   Weissman, M. M.; Leaf, P. J. *Psychol. Med.* **1988**, *18*, 141–153.

3.   Sonawalla, S. B.; Fava, M. *CNS Drugs*, **2001**, *15*, 765–776.

4.   Schildkraut, J. J. *Am. J. Psychiatry*, **1965**, *122*, 509–522.

5.   Bunney, W. E.; Davis, J. M. *Arch. Gen. Psychiatry*, **1965**, *13*, 483–494.

6.   Sneader, W. *Drug Discovery: the evolution of modern medicines.* (**1985**), pp 185–190, Wiley, Chichester.

7.   Spɪnks, D.; Spinks, G. *Curr. Med. Chem.* **2002**, *9*, 799–810.

8.   Evrard, D. A.; Harrison, B. L. *Ann. Rep. Med. Chem.* **1999**, *34*, 1–9.

9.   Feighner, J. P. *J. Clin. Psychiatry*, **1999**, *60* (*Suppl 22*), 18–22.

10.   Robertson, D. W.; Krushinski, J. H.; Fuller, R. W.; Leander, J. D. *J. Med. Chem.* **1988**, *31*, 1412–1417.

11.   Hilborn, J. W.; Lu, Z-H.; Jurgens, A. R.; Fang, Q. K.; Byers, P.; Wald, S. A.; Senanayake, C. H. *Tetrahedron Lett.* **2001**, *42*, 8919–8921.

12.   Molloy, B. B.; Schmiegel, K. K. US 4314081 (**1982**).

13.   Srebnik, M.; Ramachandran, P. V.; Brown, H. C. *J. Org. Chem.* **1988**, *53*, 2916–2920.

14.   Koenig, T. M.; Mitchell, D. *Tetrahedron Lett.* **1994**, *35*, 1339–1342.

15.   Corey, E. J.; Reichard, G. A. *Tetrahedron Lett.* **1989**, *30*, 5207–5210.

16.   Sakuraba, S.; Achiwa, K. *Chem. Pharm. Bull.* **1995**, *43*, 748–753.

17.   Kumar, A.; Ner, D. H.; Dɪke, S. Y. *Tetrahedron Lett.* **1991**, *32*, 1901–1904.

18.   Chenevert, R.; Fortier, G.; Bel Rhlɪd, R. *Tetrahedron*, **1992**, *48*, 6769–6776.

19.   Schneider, M. P.; Goergens, U., *Tetrahedron Asymmetry*, **1992**, *3*, 525–528.

20.   Bracher, F.; Litz, T. *Bioorg. Med. Chem. Lett.* **1996**, *4*, 877–880.

21.   Liu, H-L.; Hoff, B. H.; Anthonsen, T. *J. Chem. Soc., Perkin Trans.* 1, **2000**, 1767–1769.

22.   Kamal, A.; Ramesh Khanna, G. B.; Ramu, R. *Tetrahedron Asymmetry*, **2002**, *13*, 2039–2051.

23.   Ali, I. S.; Sudalai, A. *Tetrahedron Lett.* **2002**, *43*, 5435–5436.

24.   Pandey, R. K.; Fernandes, R. A.; Kumar, P. *Tetrahedron Lett.* **2002**, *43*, 4425–4426.

25.   Miles, W. H.; Fialcowitz, E. J.; Halstead, E. S. *Tetrahedron*, **2001**, *57*, 9925–9929.

26.   Trost, B. M.; Fraisse, P. L.; Ball, Z. T. *Angew. Chem. Int. Ed.* **2002**, *41*, 1059–1061.

27.   Gao, Y.; Sharpless, K. B. *J. Org. Chem.* **1988**, *53*, 4081–4084.

28.   Devine, P. N.; Heid, R. M.; Tschaen, D. M. *Tetrahedron*, **1997**, *53*, 6739–6746.

29.   Williams, M.; Quallich, G. *Chem. &Ind.* **1990**, 315–319.

30.   Welch, W. M. *Adv. Med. Chem.* **1995**, *3*, 113–148.

31.   Welch, W. M.; Kraska, A. R.; Sarges, R.; Koe, K. B. *J. Med. Chem.* **1984**, *271*, 1508–1515.

32.   Welch, W. M.; Harbert, C. A.; Koe, K. B.; Kraska, A. R. US 4536518 (**1985**).

33.   Quallich, G. J.; Williams, M. T.; Friedman, R. C. *J. Org. Chem.* **1990**, *55*, 4971–4973.

34.   Quallich, G. J.; Woodall, T. M. *Tetrahedron*, **1992**, *48*, 10239–10248.

35.   Davies, H. M. L.; Stafford, D. G.; Hansen, T. *Org. Lett.* **1999**, *1*, 233–236.

36.   Davies, H. M. L.; Walji, A. M. *Org. Lett.* **2003**, *5*, 479–482.

37.   Corey, E. J.; Gant, T. G. *Tetrahedron Lett.* **1994**, *30*, 5373–5376.

38.   Lautens, M.; Rovis, T. *J. Org. Chem.* **1997**, *62*, 5246–5247.

39.   Lautens, M.; Rovis, T. *Tetrahedron*, **1999**, *55*, 8967–8976.

40.   Christensen, J. A.; Squires, R. F. US 4007196 (**1977**).

41.   Engelstoft, M.; Hansen, J. B. *Acta Chem. Scand.* **1996**, *50*, 164–169.

42.   Herdeis, C.; Kaschinski, C.; Karia, R.; Lotter, H. *Tetrahedron Asymmetry*, **1996**, *7*, 867–884.

43.   Amat, M.; Bosch, J.; Hidalgo, J.; Canto, M.; Perez, M.; Llor, N.; Molins, E.; Miravitlles, C.; Orozco, M.; Luque, J. *J. Org. Chem.* **2000**, *65*, 3074–3084.

44. Yu, M. S.; Lantos, I.; Peng, Z-Q.; Yu, J.; Cacchio, T. *Tetrahedron Lett.* **2000**, *41*, 5647–5651.

45. deGonzalo, G.; Brieva, R.; Sanchez, V. M.; Bayod, M.; Gotor, V. *J. Org. Chem.* **2003**, *68*, 3333–3336.

46. Liu, L. T.; Hong, P-C.; Huang, H-L.; Chen, S-F.; Wang, C-L. J.; Wen, Y-S. *Tetrahedron Asymmetry*, **2001**, *12*, 419–426.

47. Greenhalgh, D. A.; Simpkins, N. S. *Synthesis Lett.* **2002**, 2074–2076.

48. Cossy, J.; Mirguet, O.; Pardo, D. G.; Desmurs, J-R. *Tetrahedron Lett.* **2001**, *42*, 5705–5707.

49. Johnson, T. A.; Curtis, M. D.; Beak, P. *J. Am. Chem. Soc.* **2001**, *123*, 1004–1005.

Chapter 11. Anti-obesity: Orlistat (Xenical®)

USAN: Orlistat
Trade Name: Xenical®
Hoffmann-La Roche
Launched: 1999
M.W. 523.79

---

§11.1 Introduction[1–10]

Obesity is the second leading cause of death in the United States, behind only smoking. An individual with body mass index (BMI) greater than or equal to 30 kg/m² is considered obese, whereas an individual with body mass index (BMI) ranging from 27 to 30 kg/m² is overweight. Obesity is implicated with a gamut of health issues including diabetes (80% of type II diabetics are obese), hypertension, dyslipidemia, gallstones and respiratory diseases. Nowadays, obesity is increasingly becoming a serious socio-economical problem. The estimated direct annual health cost associated with obesity is $70 billion, while the total overall cost to the United States economy has been estimated to be over $140 billion. In the United States, more than 50% of the adult population is overweight, and almost 1/4 of the population is considered to be obese (BMI greater than or equal to 30). The incidence of obesity is increasing in the U.S. at a 3% cumulative annual growth rate. While the vast majority of obesity occurs in the US and Europe, the prevalence of obesity is also increasing in Japan. The prevalence of obesity in adults is 10%–25% in most countries of Western Europe.

Historically, thyroid hormone was inappropriately prescribed for weight loss. Fenfluramine hydrochloride (2, Pondimin®) by Wyeth was approved by the FDA in 1973, whereas dexfenfluramine hydrochloride (3, Redux®) also by Wyeth was approved by the FDA in 1976 as appetite depressants. The major anti-obesity effects of 2 are produced from the (R)-stereoisomer, which is a serotonin reuptake inhibitor and releasing agent. Its mechanism of action (MOA) is through modulation of the serotonin level allowing the patient to achieve a sense of satiety without actually consuming a large amount of food. In 1972, phentermine hydrochloride (3, Fastin®), also by Wyeth, was approved by the FDA as an anorexic. Phentermine (3) functions as an andrenergic agent. None of these aforementioned drugs (2–4) gained tremendous notoriety until 1992, when a group at University of Rochester led by Michael Weintraub reported[1–2] that the combination of fenfluramine (2) and phentermine (3) with different mechanisms of action significantly helped weight-loss. After the publication of that study, between 1994 and 1997, the off-label use of fenfluramine (2) with phentermine (3), popularly known as Fen–Phen, became widely over-prescribed. Unfortunately, Fen–Phen was found to cause regurgitant heart valve damage and primary pulmonary hypertension (PPH). As a consequence, the manufacturer voluntarily withdrew those drugs off the market in 1997. Currently, one anoretic drug that is on the market is Abbott's sibutramine (4, Meridia®), whose MOA is also through serotonin and noradrealine reuptake inhibition (SNRI). It was approved by FDA in 1997 and million of prescriptions have been written since then.

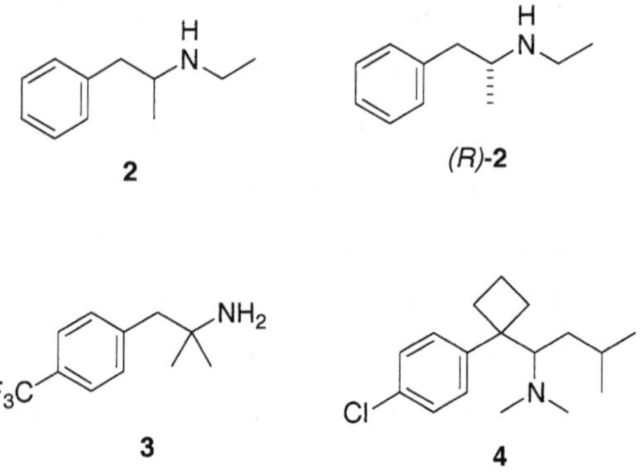

Fig. 1.

In 1987, scientists at Hoffmann-La Roche described their discovery of lipstatin (5), a secondary metabolite isolated from *Strptomyces toxytricini*, and demonstrated that it is a potent inhibitor of pancreatic lipase.[3,4] Simple hydrogenation of 5 formed terahydrolipstatin (1,USAN, orlistat) which possesses comparable biological activity but is more stable than 5.[5,6] Orlistat (1, Xenical®) works through pancreatic lipase inhibition. Pancreatic lipase is the key enzyme of dietary triglyceride absorption, exerting its activity at the water-lipid interphase, in conjunction with bile salts and co-lipase. Xenical® (1) was approved in June 1999. Millions of prescriptions have been written for this agent since launch. The β-lactone moiety was found to be required for the pancreatic lipase inhibitory activities as the ring-opened product was essentially void of the activities.

The mechanism of action (MOA) of orlistat (**1**) hinged upon the β-lactone moiety, which forms a covalent bond to the nucleophilic hydroxyl group of serine on pancreatic lipase. Conventional weight-reducing strategies have focused largely on controlling the energy intake, but there is doubt as to the long-term efficacy of these approaches. Reducing the absorption of dietary fat by lipase inhibition using Xenical® (**1**) holds great promise as an antiobesity strategy — covalent inhibition of human pancreatic lipase (HPL).[7–11]

§11.2 Synthesis of orlistat[12–20]

One of the early syntheses of orlistat (**1**) by Hoffmann-La Roche utilized the Mukaiyama aldol reaction as the key convergent step.[12,13] Therefore, in the presence of TiCl$_4$, aldehyde **7** was condensed with ketene silyl acetal **8** containing a chiral auxiliary to assemble ester **9** as the major diastereomer in a 3:1 ratio. After removal of the amino alcohol chiral auxiliary via hydrolysis, the α-hydroxyl acid **10** was converted to β-lactone **11** through the intermediacy of the mixed anhydride. The benzyl ether on **11** was unmasked via hydrogenation and the (S)-N-formylleucine side-chain was installed using the Mitsunobu conditions to fashion orlistat (**1**).

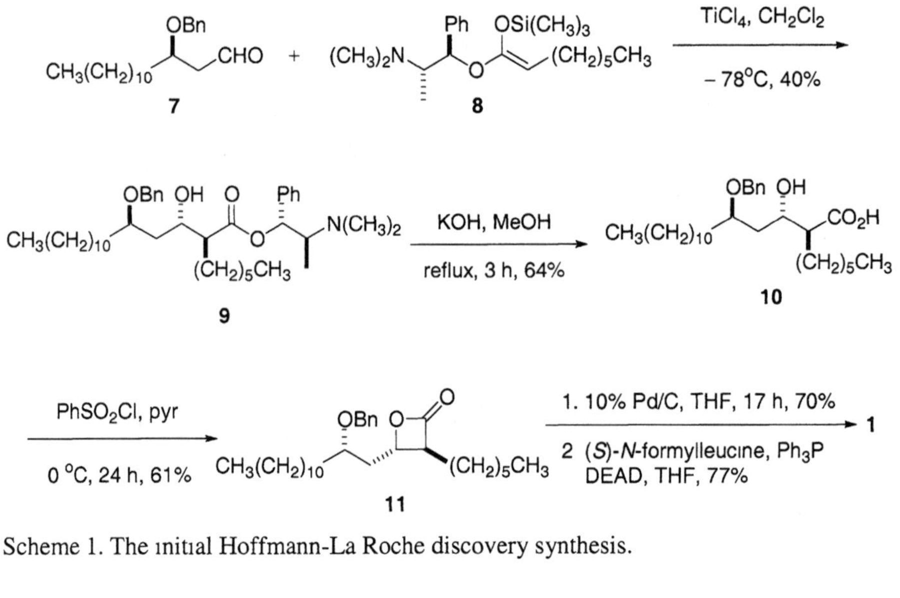

Scheme 1. The initial Hoffmann-La Roche discovery synthesis.

In 1988, an improved synthesis of orlistat (**1**) was reported by the Hoffmann-La
Roche discovery chemistry.[14]   The scheme involved a pivotal β-lactone **14**. In the
approach, an aldol condensation of aldehyde **7** with the dianion generated from octanoic
acid and two equivalents of LDA. After tosylic acid-facilitated lactonization and Jones
oxidation, the resultant lactone **14/14'** was hydrogenated to establish two additional chiral
centers. A battery of somewhat tedious protections and deprotections transformed **15** to
β-lactone **19** via the intermediacy of **16**, **17**, and **18**. Six additional steps then converted
β-lactone **19** to orlistat (**1**). This route may provide better overall yield in comparison to
the previous scheme. However, too many protections and deprotections render this
approach less elegant and not very practical for large-scale process.

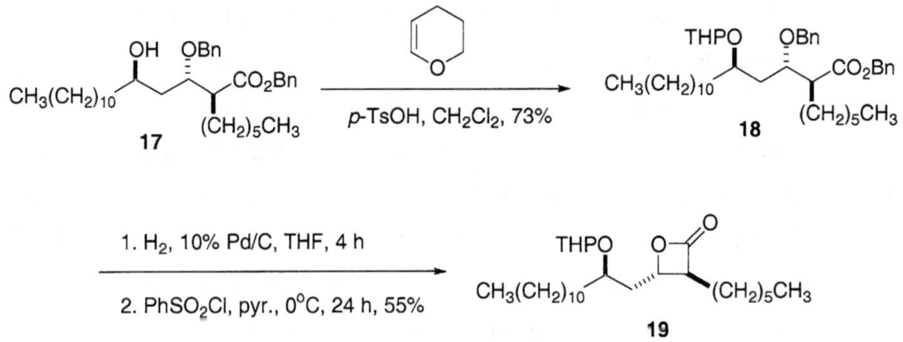

Scheme 2. The second Hoffmann-La Roche synthesis.

As the search for a better route for the orlistat (**1**) synthesis continued, Hoffmann-La Roche published a process with a better overall yield in 1991.[15] The sodium anion of cyclopentadiene was alkylated with hexyl iodide and the product underwent an asymmetric hydroboration using (+)-diisopinocampheyborane to give, after hydrogen peroxide oxidation, alcohol **20** in 65% yield and 96% *ee*. The Mitsunobu reaction of **20** with benzoic acid inverted the configuration of the alcohol. Hydrolysis of the benzoate gave rise to the corresponding diastereomer **21**. Subsequently, a hydroxyl group-directed *m*-CPBA oxidation of **21** provided the corresponding epoxide, which was protection as silyl ether **22**. Due to the steric hindrance imparted by the hexyl group, a regioselective organocuprate addition of the undecanyl group gave the epoxide ring opened product **23**. A Swern oxidation of **23** was followed by desilylation to afford ketone **24**. Very fortunately, the Baeyer–Villiger oxidation of **24** produced a single δ-lactone **25**. A string of protection and deprotection maneuvers of **25** gave α-hydroxyl acid **27** via the intermediacy of **26**. Upon exposure of **26** with benzenesulfonyl chloride, β-lactone was formed, which was hydrogenolyzed to give **27**. Conventional manipulations of **27** delivered orlistat (**1**).

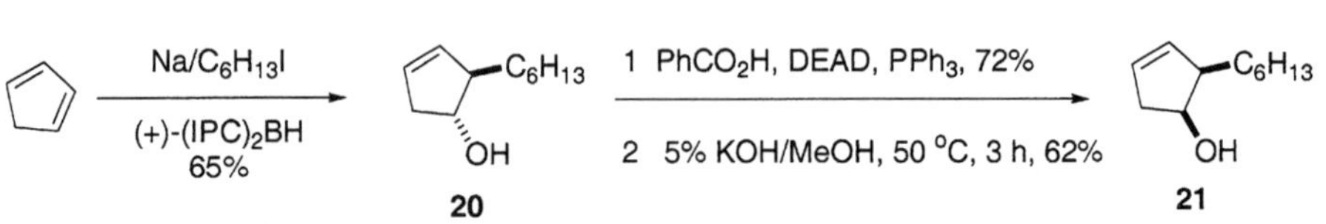

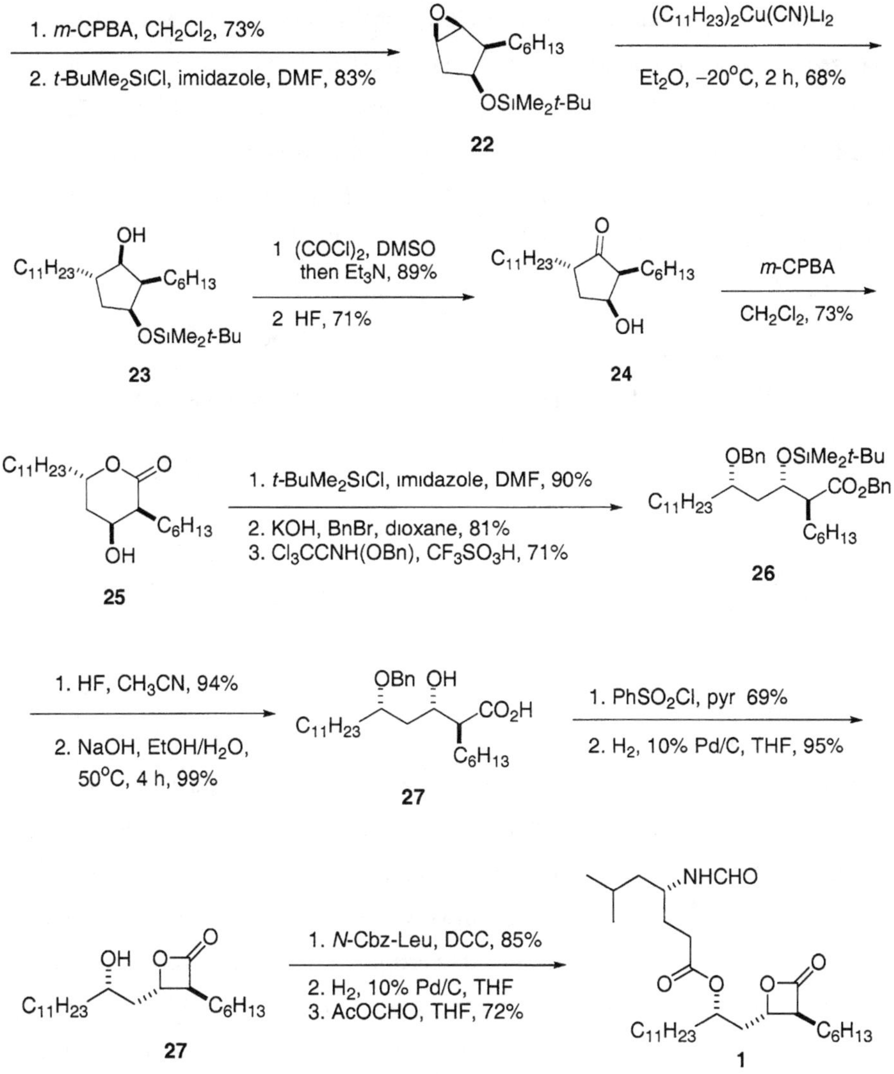

Scheme 3. The third Hoffmann-La Roche synthesis.

Thanks to the biological activities and its interesting β-lactone moiety, the synthesis of orlistat (**1**) also attracted a significant amount of attention from academia. More than ten papers have been published since 1989.[16-25] Herein only a few key examples involving novel methodology are highlighted.

Kocieński's group assembled the β-lactone segment utilizing a Lewis acid-catalyzed [2 + 2] cycloaddition strategy.[16] In the presence of a catalytic amount of boron trifluoride etherate, the [2 + 2] cycloaddition between aldehyde **28** and trimethylsilylketene **29** took place rapidly and cleanly to give a mixture of diastereomers of β-lactone **30**. After a delicate desilylation and a flash chromatography, the desired diastereomer **31** was obtained in 55% yield.

On the other hand, the Hanessian group carried out the key operation in their synthesis of orlistat (**1**) using a Lewis acid-promoted allylsilane addition to aldehyde **27**.[19] Thus, in the presence of titanium tetrachloride, aldehydes **28** and (E)-(trimethylsilyl)-2-nonene gave a mixture of diastereomers **33** as major product in an approximately 1:1 ratio. The two diasteromers were separated via flash chromatography and carried on separately. The undesired diastereomer was converted to the correct one by taking advantage of a tactic involving an enolate formation using LDA followed by acetic acid quench. Evidently, this is not a very efficient route due to the lack of diastereoselectivity of the key operation.

The Ghosh group devised an ingenious approach involving a pivotal ring-closing metathesis (RCM) step.[23] As shown below, ester-diene **34** was easily assembled from the corresponding alcohol and acryloyl chloride. The ring-closing metathesis was accomplished using Grubbs' catalyst to produce lactone **35**. Epoxidation of **35** using alkaline hydrogen peroxide afforded **36** in good selectivity. Further conventional manipulation of epoxide then delivered orlistat (**1**) via the intermediacy of diol **37**.

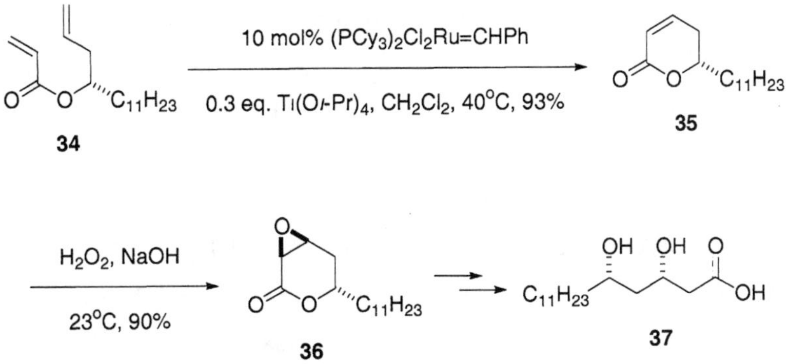

Most recently in 2003, McLeod's group obtained the β-lactone motif of orlistat (**1**) using a bromolactonization reaction.[25] β,γ-Unsaturated acid **38** was prepared from a diene-ester, hexadeca-2,4-dienoic acid methyl ester. McLeod *et al* discovered that conducting the bromolactonization in methanol and 10% aqueous sodium bicarbonate solution afforded predominantly *trans*-**39**, which was then transformed to orlistat (**1**).

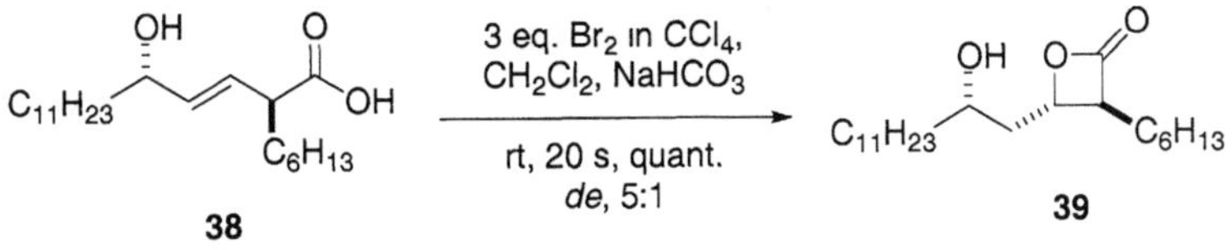

§11.3 References

1.   Weintraub, M. *Clin. Pharm.* **1992**, *51*, 581–641.

2.   Weintraub, M., Hasday, J. D., Mushlin, A. I., Lockwood, D. H. A. *Arch. Intern. Med.* **1984**, *144*, 1143–1148.

3.   Weibel, E. K.; Hadvary, P.; Hochuli, E.; Kupfer, E.; Lengsfeld, H. *J. Antibiot.* **1987**, *40*, 1081–1085.

4.   Hochuli, E.; Kupfer, E.; Maurer, R.; Meister, W.; Mercadal, Y.; Schmidt, K. *J. Antibiot.* **1987**, *40*, 1086–1091.

5.   Hadvary, P.; Hochuli, E.; Kupfer, E.; Longsfeld, H.; Weibel, E. K.; Lederer, F. EP 0129748 (**1984**).

6.   Barbier, P.; Lederer, F. EP 0189577 (**1985**).

7.   Prous, J.; Mealy, N.; Castañer, J. *Drugs Fut.* **1994**, *19*, 1003–1010.

8.   Pharmacology, McNeely, W.; Benfield, P. *Drugs* **1998**, *56*, 241–249.

9.   MOA, Borgström, B. *Biochim. Biophys. Acta* **1988**, *962*, 308–316.

10.   Clinical trial, Davidson, M. H.; Auptman, J.; DiGirolamo, M.; Foreyt, J. P.; Halsted, C. H.; Heber, D.; Heimburger, D. C.; Lucas, C. P.; Robbins, D. C.; Chung, J.; Heymsfield, S. B. *J. Am. Med. Assoc.* **1999**, *281*, 235–242.

11.   Long-term efficacy and tolerability, Hauptman, J.; Lucas, C.; Boldrin; M. N.; Collins, H.; Segal, K. R. *Arch. Farm. Med.* **2000**, *9*, 160–167.

12.   Barbier, P. Schneider, F. *Helv. Chim. Acta* **1987**, *70*, 196–202.

13.   Barbier, P.; Widmer, U.; Schneider, F. *Helv. Chim. Acta* **1987**, *70*, 1412–1418.

14.   Barbier, P.; Schneider, F. *J. Org. Chem.* **1988**, *53*, 1218–1221.

15.   Chadka, N. K.; Batcho, A. D.; Tang P. C.; Courtney, L. F.; Cook C. M.; Wovliulich, P. M.; Usković, M. R. *J. Org. Chem.* **1991**, *56*, 4714–4718.

16.   Pons, J.-M.; Kocieński, P. *Tetrahedron Lett.* **1989**, *30*, 1833–1836.

17.   Fleming, I.; Lawrence, N. J. *Tetrahedron Lett.* **1990**, *31*, 3645–3648.

18.   Case-Green, S. C.; Davies, S. G.; Hedgecock, C. J. R. *Synlett* **1991**, 781–782.

19.   Hanessian, S.; Tehim, A.; Chen, P. *J. Org. Chem.* **1993**, *58*, 7768–7781.

20.   Ghosh, A. K.; Liu, C. *Chem. Comm.* **1999**, 1743–1744.

21.   Dırat, O.; Kouklovsky, C.; Langlois, Y. *Org. Lett.* **1999**, *1*, 753–755.

22.   Welder, C.; Costisella, B.; Schick, H. *J. Org. Chem.* **1999**, *64*, 5301–5303.

23.   Ghosh, A. K.; Fidanze, S. *Org. Lett.* **2000**, *2*, 2405–2407.

24.   Schuhr, C. A.; Eisenreich, W.; Goese, M.; Stohler, P.; Weber, W.; Kupfer, E.; Bacher, A. *J. Org. Chem.* **2002**, *67*, 2257–2262.

25.   Bodkin, J. A.; Humphries, E. J.; McLeod, M. D. *Tetrahedron Lett.* **2003**, *44*, 2867–2872.

Chapter 12. Triptans for Migraine

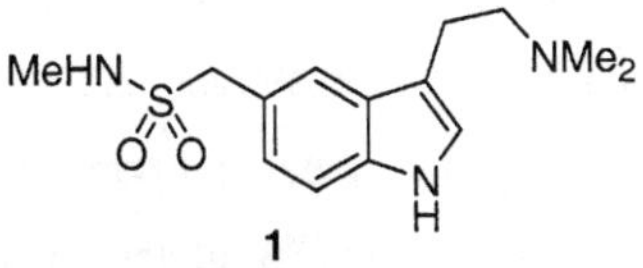

1

USAN: Sumatriptan succinate
Trade Name Imitrex®
Company: GlaxoSmithKline
Launched: 1995
M.W. 295.40

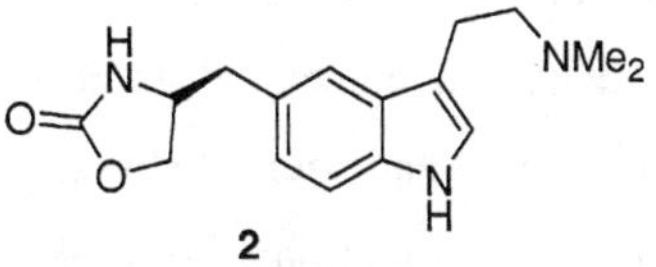

2

USAN: Zolmitriptan
Trade Name: Zomig®
Company: Wellcome/AstraZeneca
Launched: 1997
M.W. 287.36

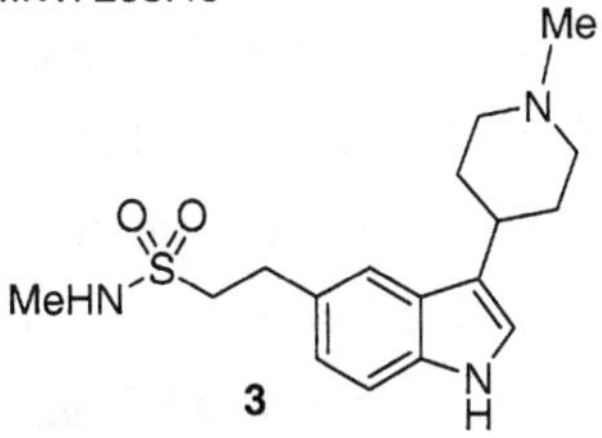

3

USAN: Naratriptan hydrochloride
Trade Name: Amerge®
Company. GlaxoSmithKline
Launched: 1998
M.W 335.47

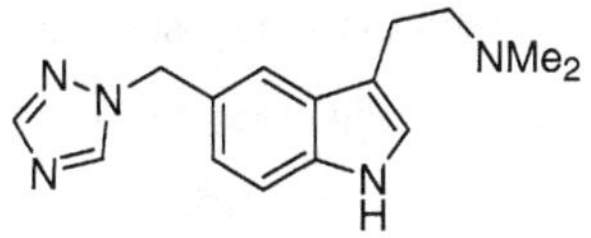

4

USAN: Rizatriptan benzoate
Trade Name: Maxalt®
Company: Merck
Launched: 1998
M W 269.35

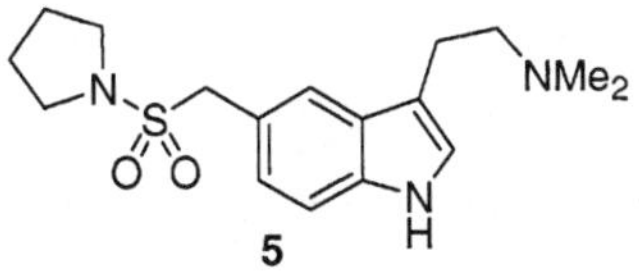

5

USAN: Almotriptan malate
Trade Name: Axert®
Company: Almirall/Janssen
Launched: 2001
M.W. 335.47

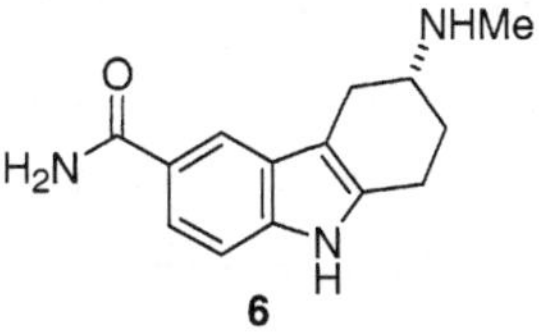

6

USAN: Frovatriptan succinate
Trade Name: Frova®
Company: SmithKline/Elan
Launched: 2001
M.W. 243.30

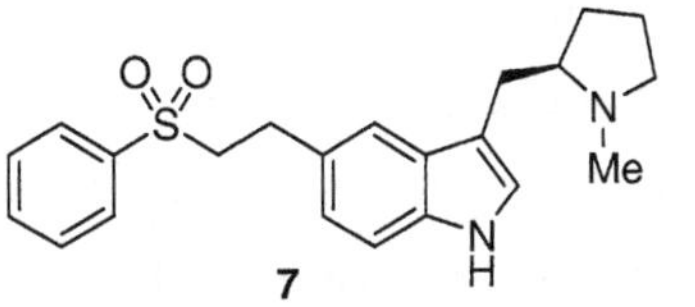

7

USAN: Eletriptan hydrobromide
Trade Name: Relpax®
Company Pfizer
Launched: 2002
M W 382 52

§12.1 Introduction[1–6]

Migraine headache is commonly characterized by intense, throbbing pain that is often aggravated by movement. Associated symptoms may include nausea and increased sensitivity to light and sound. Migraines are very common in the general population and affect 18% of women and 6% of men with approximately 35 attacks per year. Nonsteroidal anti-inflammatory drugs (NSAIDS) such as, aspirin, acetaminophen, ibuprofen, and naproxen may be effective in relieving mild migraine pain, but do not work in the majority of migraine patients.[1] Ergot alkaloids have been used for over a century for the treatment of migraines. Ergotamine is a powerful vasoconstrictor, however its effects are long lasting and are not specific to the cranial vessels, which leads to side-effects. Long after the discovery of the ergots, it was recognized that their beneficial effects resulted from activation of 5-HT_1-like receptors, specifically $5\text{-HT}_{1B/1D}$ receptors. This led to the development of sumatriptan (1), a selective $5\text{-HT}_{1B/1D}$ agonist, as the first specific antimigraine medication. It is believed that $5\text{-HT}_{1B/1D}$ agonists (triptans) elicit their antimigraine action by selective vasoconstriction of excessively dilated intracranial, extracerebral arteries and/or inhibiting the release of inflammatory neuropeptides from perivascular trigeminal sensory neurons.[2,3] It has been suggested that 5-HT_{1B} receptor activation results in vasoconstriction of intracranial vessels, while inhibition of neuropeptide release is mediated via the 5-HT_{1D} receptor. Selective 5-HT_{1D} agonists have recently been identified and are being studied to determine the relative importance of these receptor-mediated events on the antimigraine activity.[4]

Sumatriptan has proved to be an effective antimigraine drug, however it has several limitations including low bioavailability, short half-life, and a high headache recurrence rate. Since the introduction of sumatriptan (1), several second-generation triptans have entered the marketplace, including zolmitriptan (2), naratriptan (3), rizatriptan (4), almotriptan (5), frovatriptan (6) and eletriptan (7).[5,6] The second-generation triptans generally have improved pharmacokinetics, higher oral bioavailability and a longer plasma half-life (Table 1). There are subtle differences with each of the triptans with respect to efficacy, speed of onset of action, duration of action, headache recurrence rate, side-effects, and convenience of administration. Therefore, the selection of the most appropriate antimigraine drug will depend on the characterization of the patient's migraine by peak intensity, time to peak intensity, extent of disability and the extent to which the patient experiences headache recurrence and triptan related side-effects.

Triptan	Oral bioavailability (%)	$t_{1/2}$ (h)	Approved doses (mg)
Sumatriptan (**1**)	15	2	25, 50, 100
Zolmitriptan (**2**)	40–48	2.5–3	2.5, 5
Naratriptan (**3**)	63–74	5–6.3	1, 2.5
Rizatriptan (**4**)	45	2–3	5, 10
Almotriptan (**5**)	80	3.2–3.7	6.25, 12.5
Frovatriptan (**6**)	24–30	25	2.5
Eletriptan (**7**)	50	3.6–5.5	20, 40, 80

Table 1. Pharmacokinetics of the triptans.[5]

§12.2 Synthesis of sumatriptan (1)[7–12]

Sumatriptan (**1**, GR-43175) was discovered at Glaxo in the United Kingdom and the synthesis is outlined in Scheme 1.[7,8] The synthesis began with hydrogenation of *N*-methyl-4-nitrobenzenemethanesulphonamide (**8**) to give **9**. Compound **9** was treated with sodium nitrite to give a diazonium salt, which was reduced with tin chloride to give the hydrazine **10**. Condensation of **10** with 4,4-dimethoxy-*N,N*-dimethylbutylamine in aqueous HCl provided the hydrazone **11a**. Compound **11a** was treated with polyphosphate ester (PPE) in refluxing chloroform to affect the Fischer indole reaction resulting in the formation of sumatriptan (**1**) in 30% yield. The mechanism of the Fischer indole reaction is believed to be isomerization of the hydrazone from an imine to an enamine followed by [3,3] sigmatropic rearrangement and ring closure with loss of ammonia to form the indole. The low yield is partially a result of the formation of by-products such as **13** and **14** due to the nucleophilic nature of the indole product. Hydrazine **10** was also condensed with several other functionalized dimethyl acetals to give hydrazones **11b-11d**. Fischer indolization of these hydrazones afforded indoles **12b-12d**, which were also converted to sumatriptan (see Schemes 3 and 4).

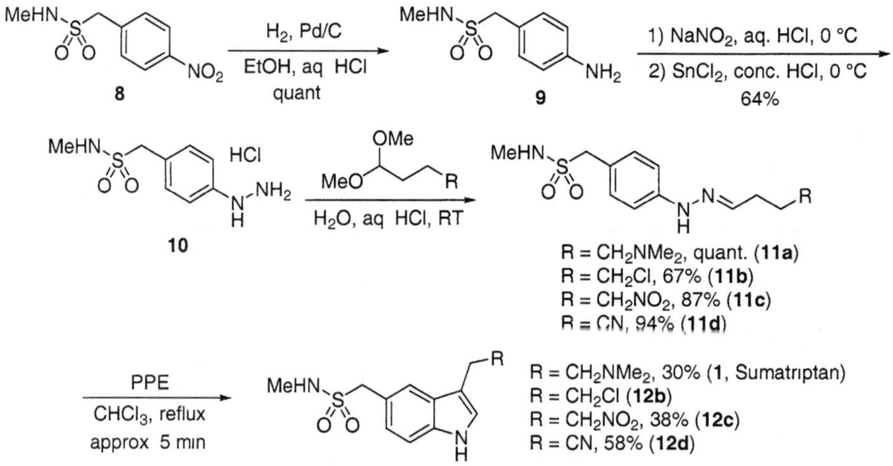

Scheme 1. The Glaxo synthesis of sumatriptan (**1**).

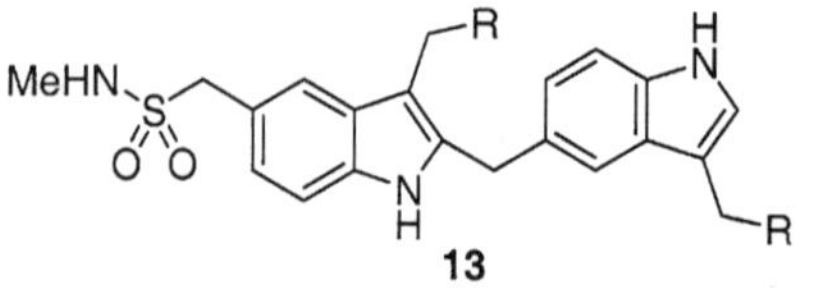

The condensation forming the hydrazone and the rearrangement can also be accomplished in a one-pot procedure as is shown in Scheme 2.[10] A mixture of *N*-protected hydrazine **15** and the acetal were treated with a catalytic amount of HCl in glacial acetic acid at 75 °C to give indole **16** in 50% yield. This convergent Fischer indole synthesis involves hydrolysis of the dimethylacetal, hydrazone formation and Fischer cyclization. By-products like **13** were avoided due to *N*-protection of the sulphonamide group, however by-product **14** was still observed. Protecting the sulphonamide group also simplified chromatographic purification. Lastly, the *N*-ethoxycarbonyl group in **16** was removed by hydrolysis to give sumatriptan (**1**).

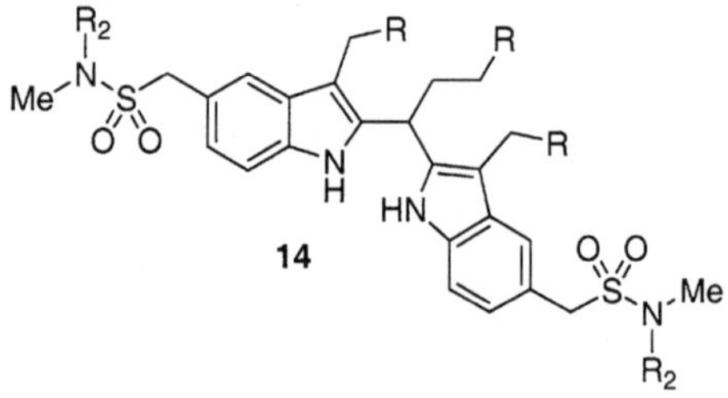

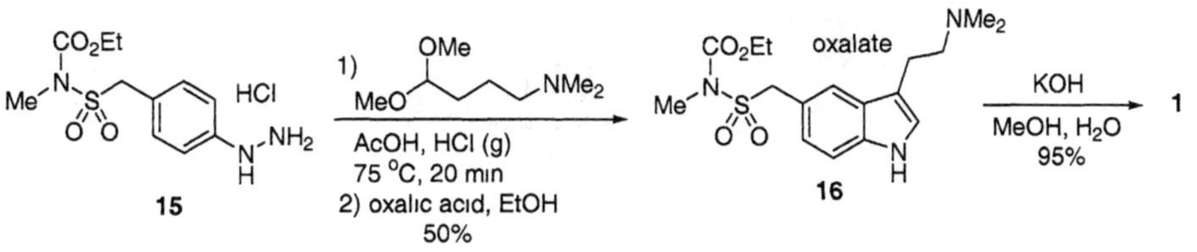

Scheme 2. Synthesis of sumatriptan (**1**) using a *N*-sulphonamide protecting group strategy.

Indole **12b**, prepared by Fischer indolization of **11b** (Scheme 1), was also converted to sumatriptan (**1**) according to Scheme 3.[7] The dimethylamino group was incorporated via a two-step procedure. First, aminolysis of **12b** with ammonia provided the aminoethyl compound **17**, albeit in low yield. Then reductive alkylation of **17** with formaldehyde in the presence of NaBH$_4$ gave sumitriptan (**1**).

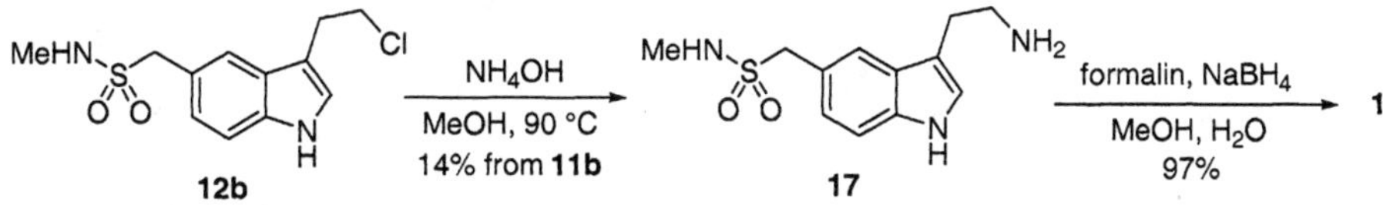

Scheme 3. The synthesis of sumatriptan (**1**) from **12b**.

2-Cyanomethylindole **12d** also proved to be a versatile intermediate toward the synthesis of sumatriptan (Scheme 4).[7,8] Hydrogenation of the cyano group of **12d** in the presence of dimethylamine provided sumitriptan (**1**) directly, which was isolated as the hemisuccinate salt. Alternatively, the cyano group of **12d** was hydrolyzed to the carboxylic acid (**18**) and then transformed into the dimethyl amide **19**. Finally, the amide was reduced with LiAlH$_4$ to afford sumatriptan (**1**). Although this is a longer synthesis of **1**, the intermediate **18** allows for the preparation of various tryptamine analogs.

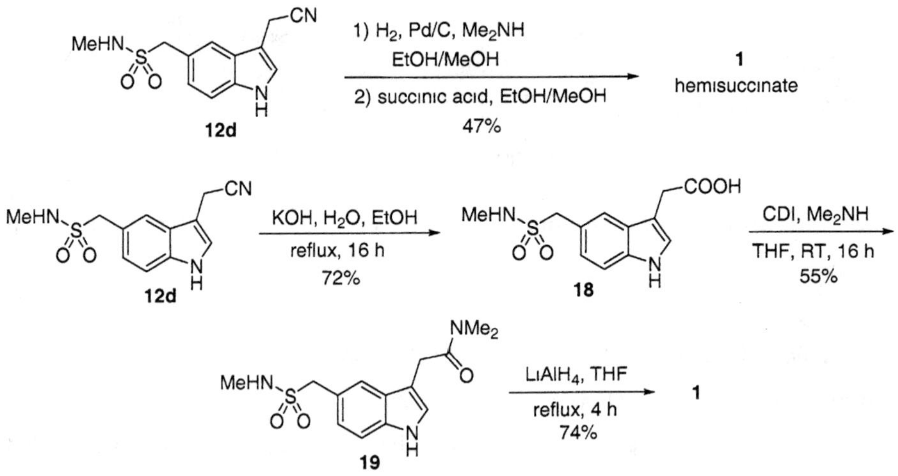

Scheme 4. Synthesis of sumatriptan (**1**) from **12d**.

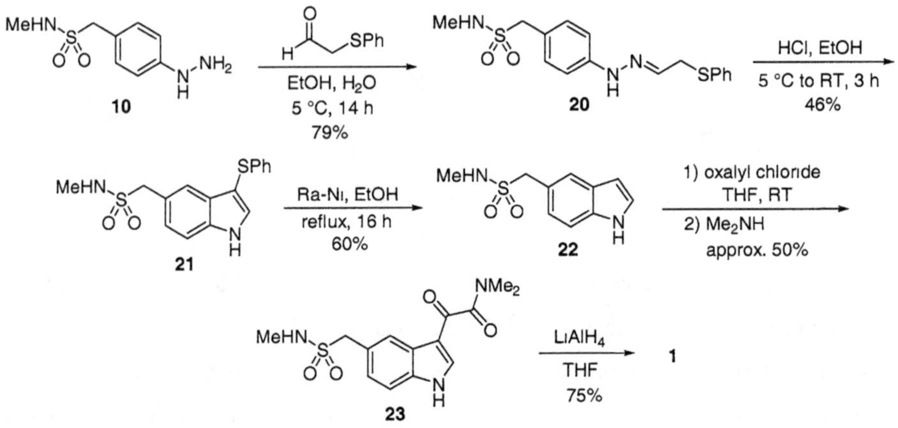

Scheme 5. An alternative synthesis of sumatriptan (**1**).

An alternative synthesis from the Glaxo patents involves Friedel-Crafts acylation of the 3-position of the indole intermediate **22** (Scheme 5).[8] Reaction of hydrazine **10** with (phenylthio)acetaldehyde gave hydrazone **20**, which was subjected to the Fischer indole reaction to give 3-thiophenylindole **21**. It is noteworthy that this Fischer cyclization took place at room temperature because most require heat. Reductive desulfurization of **21** using Raney nickel provided indole **22**. Acylation of the 3-position of the indole nucleous with oxalyl chloride followed by treatment with dimethylamine gave the α-ketoamide **23**. Finally, reduction of the dicarbonyl moiety was effected by LiAlH₄ to afford sumitriptan (**1**).

Knoll has recently published a more stream-lined and optimized synthesis of sumatriptan (**1**) using very similar methodology to that described above (Scheme 6).[9] The hydrogenation, diazonium formation and reduction of the diazonium salt to form **10** can be carried out as a one-pot reaction. In this scenario, nitro compound **8** was hydrogenated in aqueous HCl and the palladium catalyst was removed by filtration. The filtrate was cooled to –5 °C and sodium nitrate was added to form the diazonium salt. The reaction solution was then transferred to a solution of sodium dithionite in aqueous NaOH and isopropanol to give hydrazine **10**. The use of sodium dithionite is an improvement over stannous chloride since it is a cheaper reducing agent with less environmental problems. Hydrazine hydrochloride **10** was treated with 4-chlorobutanal dimethyl acetal in ethanol and aqueous HCl to give the intermediate hydrazone. Sodium hydrogenphosphate was added and the reaction mixture was refluxed to produce **17**. This reaction is known as the Grandberg version of the Fischer indole synthesis and involves displacement of the chloro group with the ammonia released during the formation of the indole ring. Reaction of tryptamine **17** with formaldehyde and NaBH₄ in the presence of a buffer provided sumatriptan (**1**). The Knoll patent states that triptans **2–5** and **7** can be prepared in a similar manner.

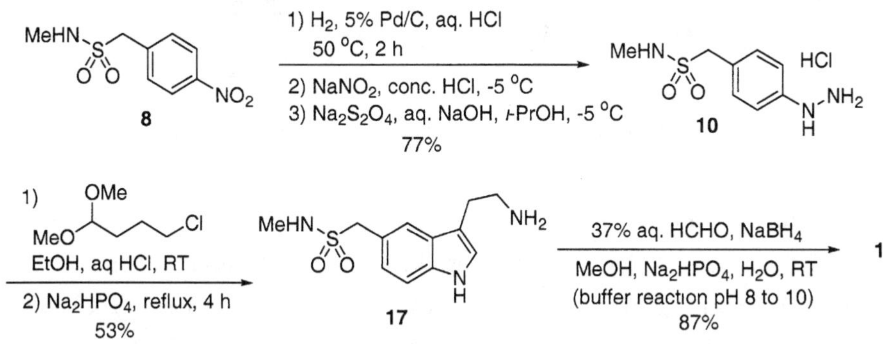

Scheme 6. Knoll synthesis of sumatriptan (**1**).

A group from Hungary has synthesized sumatriptan (**1**) utilizing a Japp-Klingemann reaction/decarboxylation strategy (Scheme 7).[11]   Compound **24** was diazotized with sodium nitrite to give the diazonium salt **25**, which underwent a Japp-Klingemann reaction with ethyl 2-(3-dimethylaminopropyl)-3-oxobutanoate in the presence of sodium acetate to give hydrazone **26**. Fischer cyclization of **26** proceeded smoothly when treated with acetic acid and HCl gas at ambient temperature. One of the advantages of this strategy is that the 2-carboethoxy substituted indole **27** is not subject to electrophilic attack and by-products like **13** and **14** can be avoided. The ethyl ester of **27** was hydrolyzed and the resulting acid **28** was decarboxylated by heating at 200 °C in quinoline with copper powder to give **1**.

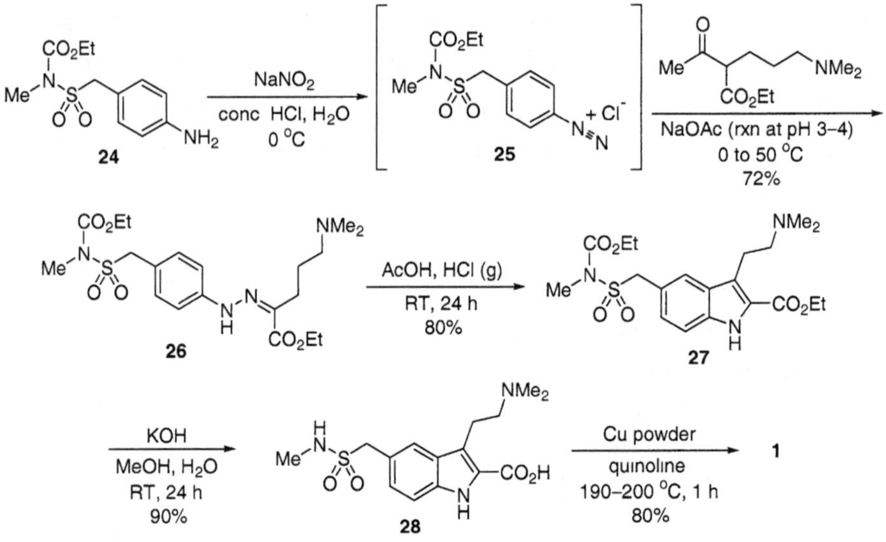

Scheme 7. Synthesis of sumatriptan (**1**) using the Japp-Klingemann reaction.

§12.3 Synthesis of zolmitriptan (2)[13–16]

Zolmitriptan (**2**, BW-311C90) was discovered by Wellcome in the United Kingdom and then transferred to AstraZeneca upon the acquisition of Wellcome by Glaxo (now GlaxoSmithKline). The synthesis began with 4-nitro-*L*-phenylalanine (**29**), which was converted to the methyl ester and then reduced with NaBH$_4$ to give the amino alcohol (Scheme 8).[13,14] Oxazolidinone **30** was formed by treating the amino alcohol with phosgene. The nitro group of **30** was hydrogenated and the resulting aniline was diazotized and reduced with stannous chloride to afford hydrazine **31**. Fischer cyclization of the hydrochloride salt of **31** with 4-chlorobutanal diethyl acetal in ethanol/water provided tryptamine **32** in low yield. Dimethylation of **32** using reductive amination conditions afforded zolmitriptan (**2**). Alternatively, Fischer cyclization of the hydrochloride salt of **31** with 4,4-diethoxy-*N,N*-dimethylbutylamine in refluxing acetic acid directly yielded zolmitriptan (**2**) (Scheme 9).[14]

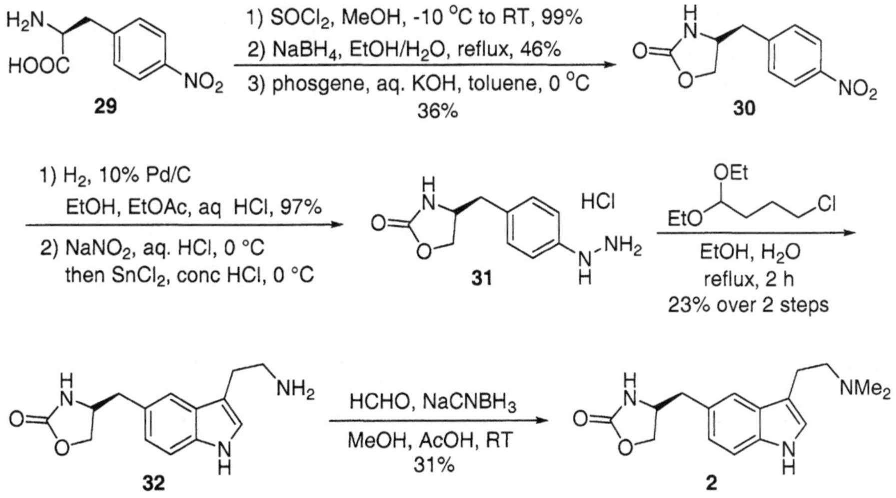

Scheme 8. The Wellcome synthesis of zolmitriptan (**2**).

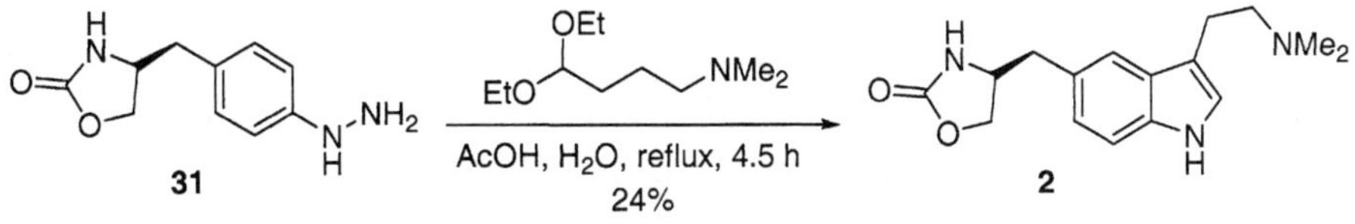

Scheme 9. An alternative synthesis of zolmitriptan (**2**).

The yield of the Fischer indole reaction was improved dramatically by employing 3-cyanopropanal diethylacetal as the "aldehyde" coupling partner (Scheme 10).[14] Thus the hydrazone **33** was formed by reacting the hydrochloride salt of **31** with the diethylacetal in aqueous HCl. The hydrazone **33** was treated with polyphosphate ester in refluxing chloroform to give the cyanomethyl indole **34** in 77% yield. Finally, hydrogenation of the cyano group of **34** in the presence of dimethylamine provided zolmitriptan (**2**) in a single step.

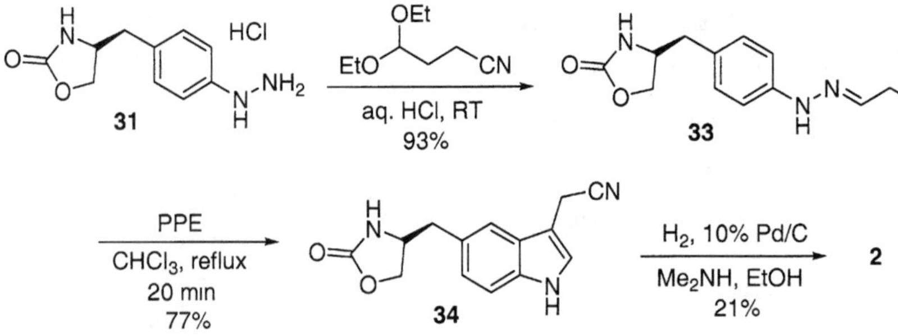

Scheme 10. Synthesis of zolmitriptan (**2**) from cyano intermediate **33**.

AstraZeneca has recently reported a "one-pot" synthesis of zolmitriptan on a multi-kilogram scale (Scheme 11).[15] Methyl 4-nitro-(L)-phenylalaninate hydrochloride (**35**) was neutralized with Na_2CO_3 and treated with *n*-butyl chloroformate. The mixture was extracted with ethyl acetate and the resulting ethyl acetate solution of the butyl carbamate product was used directly in the subsequent hydrogenation reaction. Hydrogenation of the nitro group using 5% Pd/C provided the corresponding aniline. The ethyl acetate was distilled from the reaction and replaced with butanol. The butanol solution was then treated with $NaBH_4$ to reduce the methyl ester to the corresponding alcohol. The excess $NaBH_4$ was quenched with aqueous HCl and the mixture was basified with ammonia. The butanol layer was separated and treated with sodium methoxide at 85 °C to give the oxazolidinone **36**. The reaction mixture was cooled to 5 °C to give a precipitate and **36** was isolated by filtration. Compound **36** was dissolved in aqueous HCl and treated with aqueous $NaNO_2$. The resulting diazonium salt solution was added to an aqueous solution of sodium sulphite to give the corresponding hydrazine. Next 4,4-diethoxy-*N,N*-dimethylbutylamine was added and the mixture was refluxed for 4 hours. The reaction was quenched and extracted with ethyl acetate to give the crude product, which was recrystallized from 10% ethanol in ethyl acetate to give pure zolmitriptan (**2**).

The patent does not give yields except to say that the final product (2) is formed in high yield. This synthesis only requires the isolation of one intermediate (36) and does not require any purification until the recrystallization of the final product (2). Furthermore, this route avoids the use of hazardous reagents such as phosgene and tin chloride (see Scheme 8).

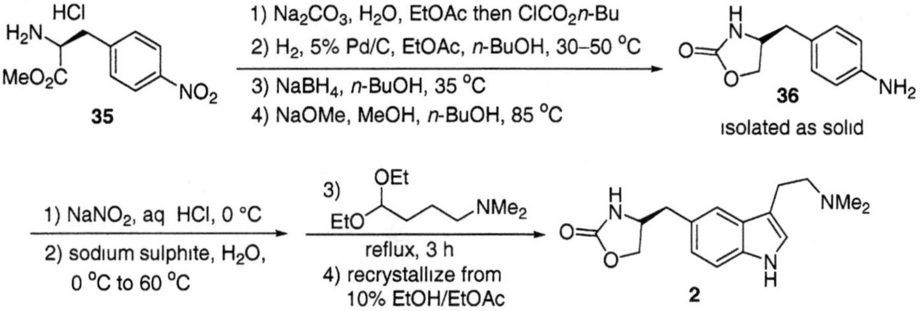

Scheme 11. AstraZeneca "one-pot" synthesis of zolmitriptan (2).

§12.4 Synthesis of naratriptan (3)[17–20]

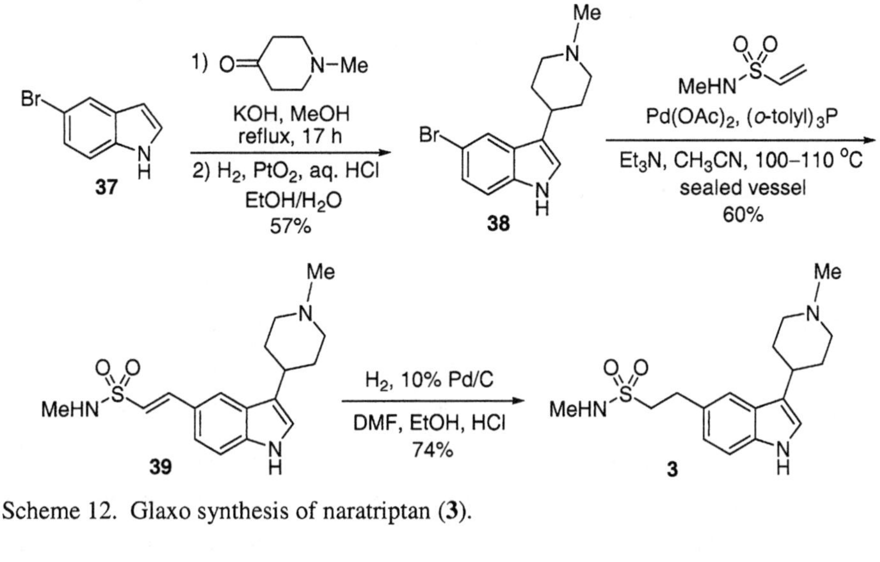

Scheme 12. Glaxo synthesis of naratriptan (3).

Naratriptan (**3**, GR-85548) was discovered at Glaxo in the United Kingdom as a follow-on to sumatriptan (**1**) and the synthesis is shown in Scheme 12.[17] Reaction of 5-bromoindole (**37**) with *N*-methyl-4-piperidone in refluxing methanolic KOH gave a *N*-methyl-tetrahydropyridinyl substituted indole, which was hydrogenated immediately using 5% platinum oxide on carbon to give *N*-methylpiperidine **38**. Heck reaction of **38** with *N*-methylvinylsulfonamide at 100–110 °C in a sealed reaction vessel provided **39**. Hydrogenation of **39** in ethanolic HCl and DMF afforded naratriptan (**3**).

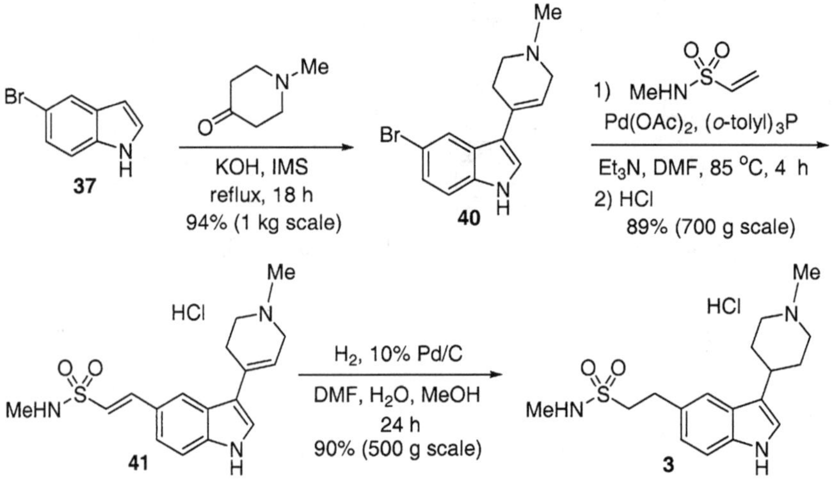

Scheme 13. Large-scale synthesis of naratriptan (**3**).

More recently, a patent application from Glaxo was published detailing a process for the synthesis of naratriptan on a large scale (Scheme 13).[18] The route is very similar to the synthesis shown in Scheme 12, except one step was removed and the other steps were optimized leading to an efficient synthesis of **3**. Instead of hydrogenating compound **40**, it was subjected to the Heck reaction with *N*-methylvinylsulfonamide and diene **41** was isolated as the hydrochloride salt in 89% yield. Both double bonds of diene **41** were hydrogenated to give naratriptan hydrochloride (**3**) as white crystals in 71% yield after recrystallization from hot water. An alternative preparation of intermediate **41** from **40** is shown in Scheme 14. Bromoindole **40** was lithiated with butyl lithium and then reacted with DMF to provide aldehyde **42**. The anion of *N*-methylmethylsulfonamide was added to aldehyde **42** and the product was dehydrated under acidic or basic conditions to give **41**. Diene **41** could then be converted to naratriptan (**3**), as shown in Scheme 13.

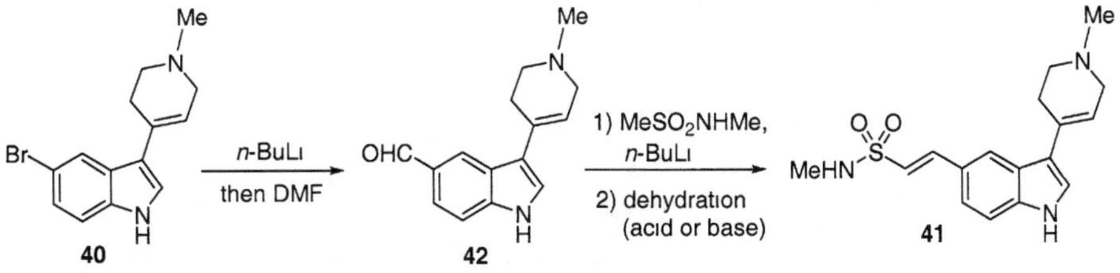

Scheme 14. Alternative preparation of intermediate **41**.

Naratriptan was also prepared using the Fischer indole reaction (Scheme 15).[17] Thus hydrazine **43** was condensed with *N*-methyl-4-piperidineacetaldehyde in aqueous HCl to form hydrazone **44**. Fischer indolization was effected by treating **44** with polyphosphate ester in refluxing CHCl₃ to provide naratriptan (**3**) in low yield. Higher yields may be attainable with the use of milder Fischer cyclization conditions (*i.e.* acetic acid or aqueous HCl).

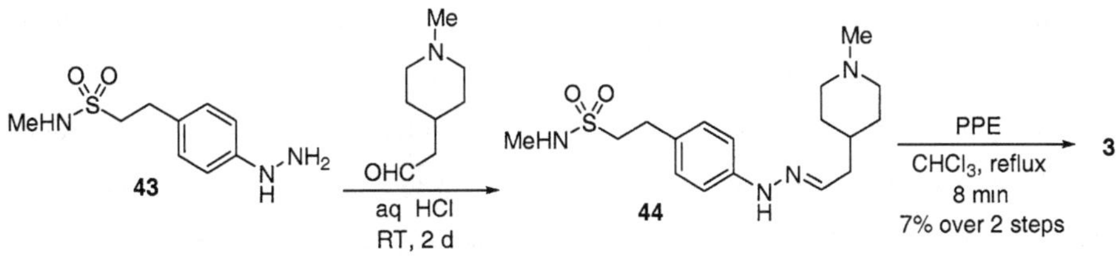

Scheme 15. Synthesis of naratriptan (**3**) using the Fischer indole reaction.

The synthesis of a ¹⁴C-labelled version of naratriptan (**3b**) is highlighted in Scheme 16.[20] The indole ring of naratriptan hydrochloride (**3**) was oxidatively cleaved using sodium periodate to give ketoformanilide **45**. Cyanation of **45** with potassium [¹⁴C]cyanide in aqueous ethanol gave the intermediate amidine **46**, which was reduced directly with NaBH₄ in acetic acid to afford ¹⁴C-labelled naratriptan (**3b**), which was isolated as the hydrochloride salt.

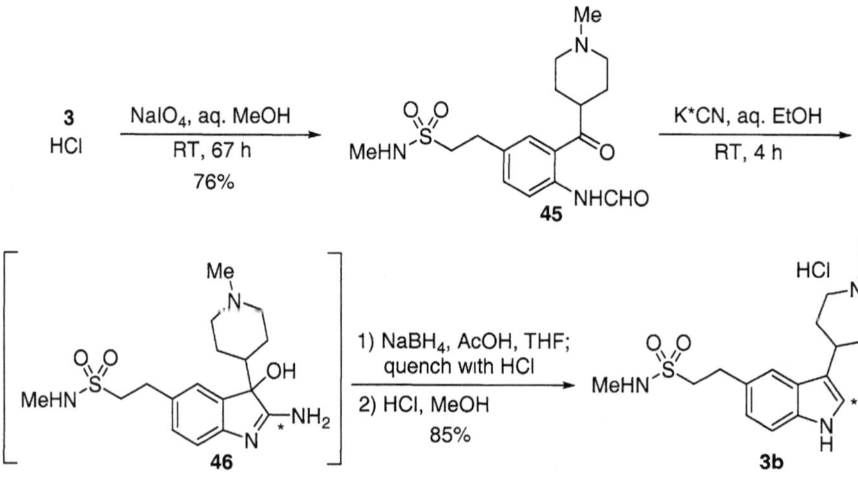

Scheme 16. Synthesis of ^{14}C-labelled naratriptan (**3b**).

§12.5 Synthesis of rizatriptan (4)[21–27]

Rizatriptan (**4**, MK-0462) was discovered by Merck in the United Kingdom. It is an analog of sumatriptan (**1**) where the *N*-methylsulfonamide group has been replaced with a 1,2,4-triazole. The synthesis begins with reaction of the sodium anion of 1,2,4-triazole with 4-nitrobenzyl bromide followed by hydrogenation of the nitro group to give **47** (Scheme 17).[21,22] Compound **47** was converted to the diazonium salt and then reduced with stannous chloride to afford hydrazine **48**. Application of the Grandberg version of the Fischer indole synthesis between hydrazine **48** and 4-chlorobutanal dimethyl acetal in ethanol and aqueous HCl provided indole **49** in a single step (also see Scheme 6). Reductive alkylation of tryptamine **49** with formaldehyde in the presence of NaCNBH$_3$ afforded rizatriptan (**4**). *N,N*-dimethyltryptamine **4** can also be formed directly from hydrazine **48** by treatment with 4-(*N,N*-dimethylamino)butanol dimethyl acetal in refluxing 4% H$_2$SO$_4$. The low yield of these Fischer indole reactions is partially due to the instability of the benzyl triazole moiety to the reaction conditions.

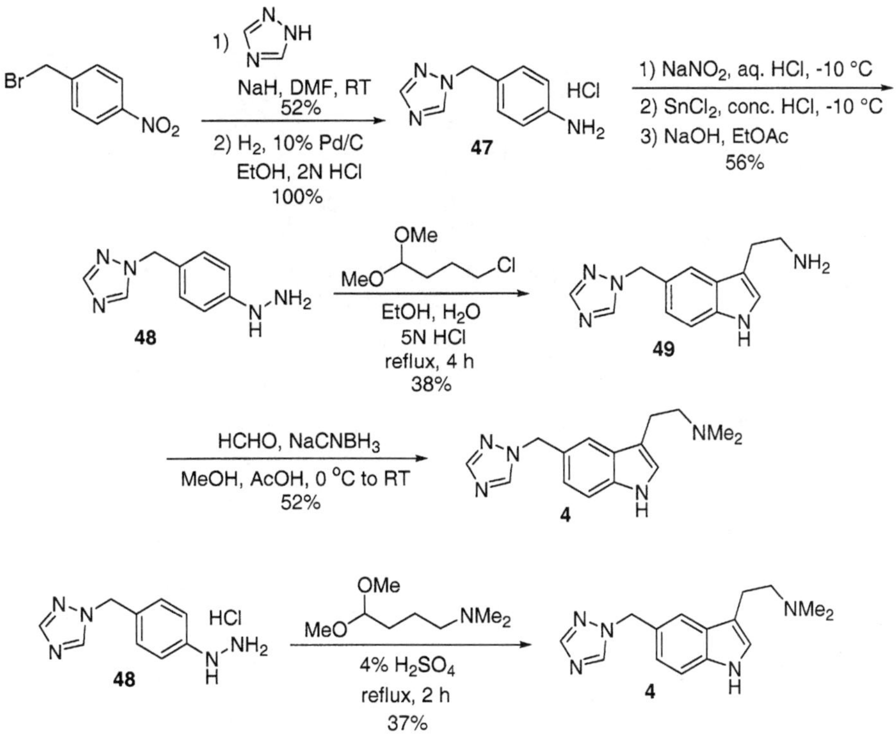

Scheme 17. Merck synthesis of rizatriptan (4).

A similar, but better yielding synthesis of **4** from Merck is presented in Scheme 18.[23] 4-Amino-1,2,4-triazole was alkylated with 4-nitrobenzyl bromide by simply refluxing the mixture in isopropanol to give **50** in excellent yield. The aminotriazole **50** was deaminated with NaNO₂ in aqueous HCl and the nitro group was reduced with ammonium formate catalyzed by 10% Pd/C to deliver **47** in an improved yield over the route shown in Scheme 17. Diazotization of **47**, reduction of the diazonium salt with sodium sulphite and Fischer indolization of the resulting hydrazine with 4-(*N,N*-dimethylamino)butanol dimethyl acetal was performed in a single step to afford rizatriptan (**4**) in 45% yield.

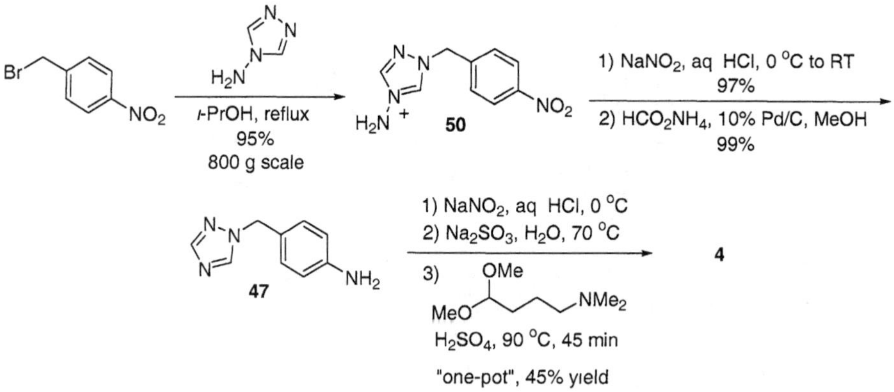

Scheme 18. Synthesis of rizatriptan (**4**) starting from 4-amino-1,2,4-triazole.

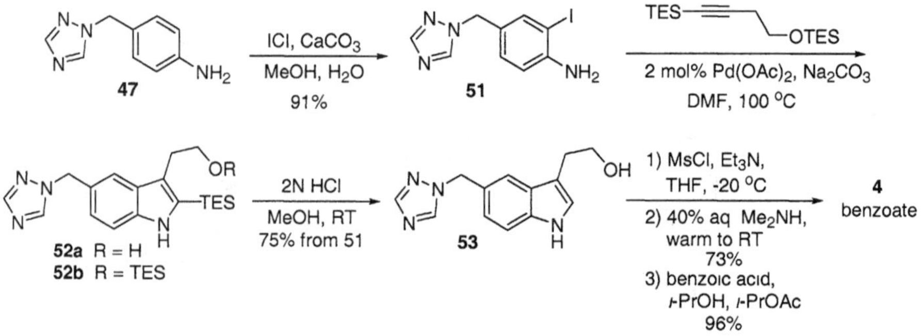

Scheme 19. Synthesis of rizatriptan (**4**) via Pd-catalyzed coupling of an iodoaniline and an alkyne.

The Merck process group in Rahway has developed two syntheses of rizatriptan (**4**) utilizing palladium catalyzed indolization reactions (Schemes 19 and 20).[24-26] Both routes start from the iodoaniline **51**, which was prepared by reaction of **47** with iodine monochloride in the presence of CaCO$_3$.[24,25] Palladium catalyzed coupling of iodoaniline **51** with bis-triethylsilyl protected butynol in the presence of Na$_2$CO$_3$ provided a mixture of indoles **52a** and **52b**. This mixture was desilylated with aqueous HCl in MeOH to furnish the tryptophol **53** in 75% yield from **51**. Protection of the alkyne prevented coupling at the terminal carbon of the alkyne and triethylsilyl (TES) was found to be optimal because it offered the correct balance between reactivity (rate of coupling) and stability. Notably, this palladium catalyzed process does not require the use of triphenyl phospine, tetrabutyl ammonium chloride or lithium chloride. The alcohol **53** was converted to the mesylate and then treated directly with 40% dimethylamine to afford **4**, which was purified by forming the benzoate salt.

The second route involves the palladium catalyzed coupling of iodoaniline **51** with the acyl silane **54** in the presence of DABCO to give the 2-trimethylsilyl indole **55**.[26] The acyl silane **54** was prepared by alkylation of 1,3-dithianyl-trimethylsilane with (3-bromopropyl)-dimethylamine to give **56**, followed by removal of the dithioacetal with mercuric oxide and mercuric chloride. Finally, desilylation of **55** in aqueous HCl and methanol afforded rizatriptan (**4**).

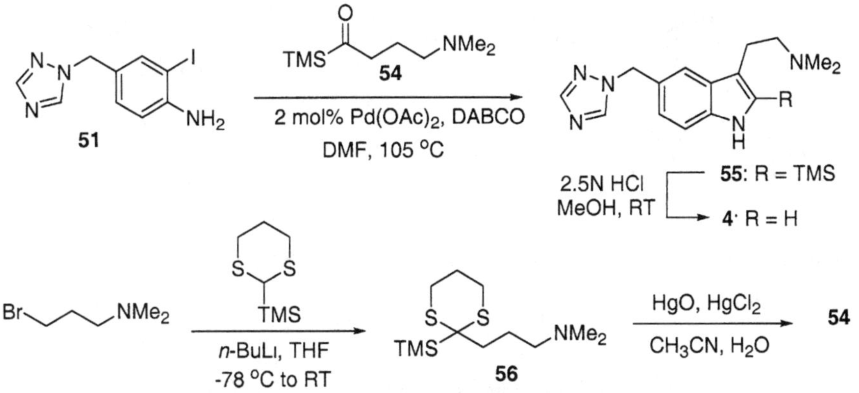

Scheme 20. Synthesis of rizatriptan (**4**) via Pd-catalyzed coupling of iodoaniline **51** and acyl silane **54**.

§12.6 Synthesis of almotriptan (5)[28–31]

Almotriptan (**5**, LAS-31416) was discovered and developed by Almirall in Spain. The U.S. marketing rights were licensed to Pharmacia and then to Janssen. The synthesis commenced with the reaction of (*p*-nitrophenyl)-methanesulfonyl chloride (**57**) with pyrrolidine, followed by hydrogenation of the nitro group to give aniline **58** (Scheme 21).[28] The hydrazine **59** was prepared by diazotization of **58** followed by reduction of the resulting diazonium salt with stannous chloride. Hydrazine **59** was treated with 4-chlorobutanal diethyl acetal in aqueous HCl and hydrazone **60** precipitated and was isolated by filtration. Treatment of the crude hydrazone **60** with aqueous HCl buffered to pH 5 with Na$_2$HPO$_4$ promoted the Grandberg modification of the Fischer indole synthesis whereby the alkyl chloride underwent aminolysis under the reaction conditions to afford the tryptamine **61** in reasonable yield. Reductive alkylation of **61** with formaldehyde in the presence of NaBH$_4$ provided almotriptan (**5**).

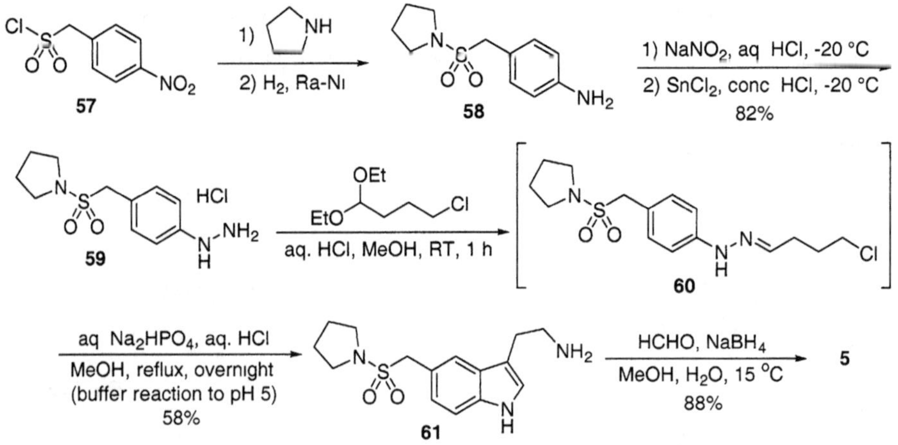

Scheme 21. Synthesis of almotriptan (**5**) using the Grandberg modification of the Fischer indole reaction.

Almotriptan has also been synthesized via decarboxylation of the carboxylic acid intermediate **65**, but a detailed preparation of **65** was not provided in the patent literature (Scheme 22).[29] The patent indicates that the carboxy indole **65** was prepared according to the method of Gonzalez.[30] Thus, (2-oxo-tetrahydro-3-furanyl)-glyoxylic acid ethyl ester (**62**) was heated in aqueous H$_2$SO$_4$ to give 2-oxo-5-hydroxypentanoic acid *in situ*, which was treated with hydrazine **59** to produce hydrazone **63**. Fischer cyclization of **63** using HCl gas in DMF gave the lactone **64**, which was converted to carboxylic acid **65**. Decarboxylation of **65** was catalyzed by cuprous oxide in quinoline at 190 °C to afford almotriptan (**5**).[29]

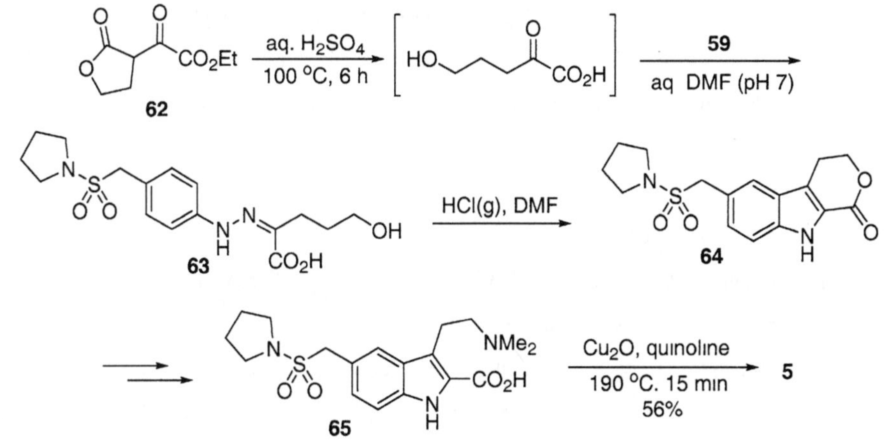

Scheme 22. Synthesis of almotriptan (**5**) employing a Fischer indolization/decarboxylation strategy.

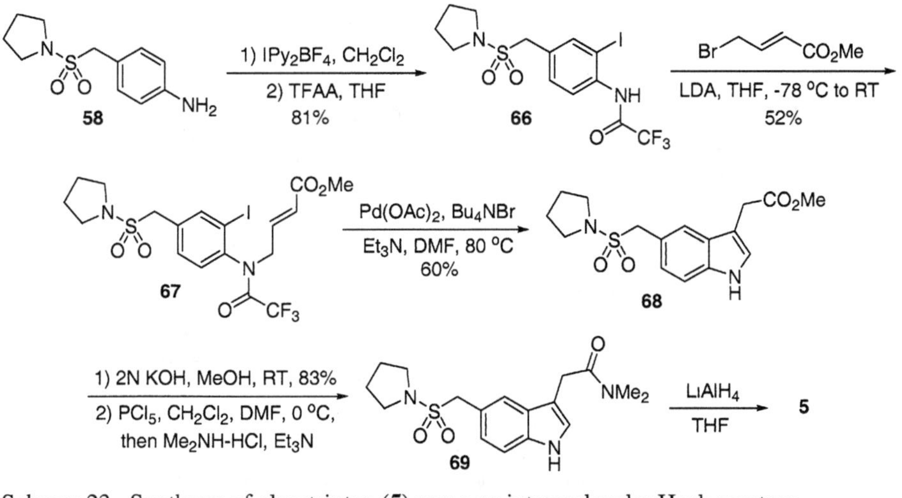

Scheme 23. Synthesis of almotriptan (**5**) using an intramolecular Heck reaction.

Lastly, a formal synthesis of almotriptan (**5**) has been published by the Bosch group in Spain using an intramolecular Heck approach (Scheme 23).[28] Iodination of aniline **58** with bis(pyridine)iodonium(I) tetrafluoroborate followed by reaction with trifluoroacetic anhydride (TFAA) gave the *o*-iodotrifluoroacetanilide **66**. Alkylation of **66** with methyl 4-bromocrotonate afforded **67**, which underwent palladium catalyzed

Heck cyclization with concomitant deprotection of the trifluoroacetyl group to form the indoleacetic methyl ester **68**. Compound **68** was hydrolyzed and then treated with PCl$_5$ followed by dimethylamine to give the amide **69**, which was reduced with LiAlH$_4$ to afford almotriptan (**5**).

§12.7 Synthesis of frovatriptan (6)[32–34]

Frovatriptan (**6**, SB-209509/VML-251), initially discovered by SmithKline Beecham, was licensed to Vanguard Medica/Vernalis for development. The marketing rights for frovatriptan have subsequently been licensed to Elan in the United States. Frovatriptan is a conformationally-restrained analog of the selective 5-HT$_1$ agonist 5-carbamoyltryptamine (5-CT). The key step in the syntheses involve Fischer reaction of 4-cyanophenylhydrazine (**70**) or 4-carboxamidophenylhydrazine (**79**) with an appropriately substituted cyclohexanone (Schemes 24–26).[32,33]

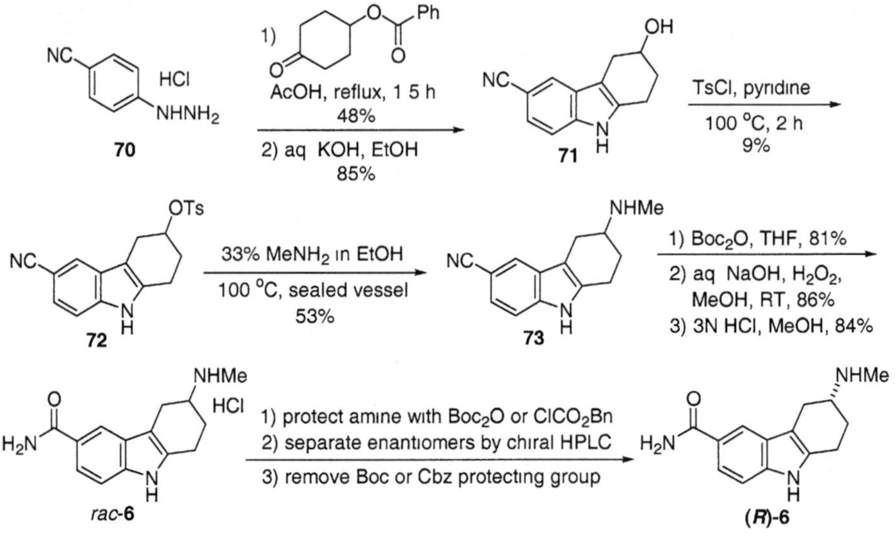

Scheme 24. Synthesis of optically pure frovatriptan ((*R*)-**6**) utilizing 4-benzoyloxycyclohexanone in the Fischer indole cyclization.

Reaction of 4-cyanophenylhydrazine hydrochloride (**70**) with 4-benzoyloxycyclohexanone in refluxing acetic acid furnished the Fischer indole product and the benzoyl group was removed by hydrolysis to give **71** (Scheme 24).[32,33] The hydroxy group of **71** was converted to the tosylate (**72**) in poor yield and then displaced with methylamine to afford **73**. Following protection of the amine with di-*tert*-butyl dicarbonate, the cyano group of **73** was converted to the carboxamide using hydrogen peroxide in the presence of sodium hydroxide. Racemic frovatriptan (*rac*-**6**) was prepared by removal of the Boc group or enantiomerically pure frovatriptan ((*R*)-**6**) was prepared by first separating the enantiomers by chiral HPLC followed by removal of the Boc protecting group.   The optical resolution can also be performed with the benzoyloxycarbonyl protected derivative of *rac*-**6**.

A more convergent synthesis of frovatriptan using the methylamino-substituted cyclohexanone equivalent **75** is shown in Scheme 25.[33] The mono-ketal of 1,4-cyclohexanedione (**74**) was treated with methylamine in ethanol and then hydrogenated to give **75** as an oil, which was converted to the hydrochloride salt. The hydrazine of **76**, formed *in situ* by treatment with sodium nitrite followed by reduction of the diazonium salt with sodium dithionite, was reacted with **75** and additional concentrated HCl at 70 °C to deliver racemic frovatriptan (*rac*-**6**).

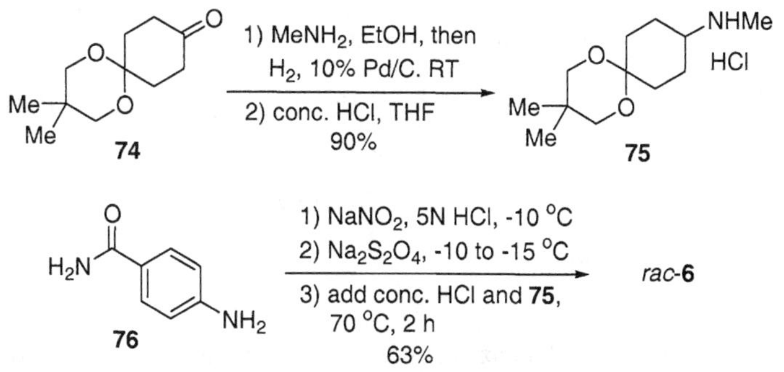

Scheme 25. Convergent "one-pot" synthesis of racemic frovatriptan from **75** and **76**.

Scheme 26 describes a convergent synthesis of des-methyl frovatriptan ((*R*)-**81**).[32,33]   4-Aminocyclohexanol   hydrochloride   was   reacted   with   N-carboethoxyphthalimide to form 4-phthalimido cyclohexanol (**77**), which was oxidized with PCC to provide cyclohexanone **78**. Fischer indolization of hydrazine **79** with **78** in refluxing acetic acid yielded **80**. The phthalimide was removed by treatment with hydrazine to furnish the versatile amine intermediate **81**. The enantiomers of **81** were resolved by recrystallization of a diastereomeric salt or by chiral HPLC separation of the *tert*-butyl carbamate of **81** to give (*R*)-**81**, which was converted to frovatriptan ((*R*)-**6**) according to Scheme 27.

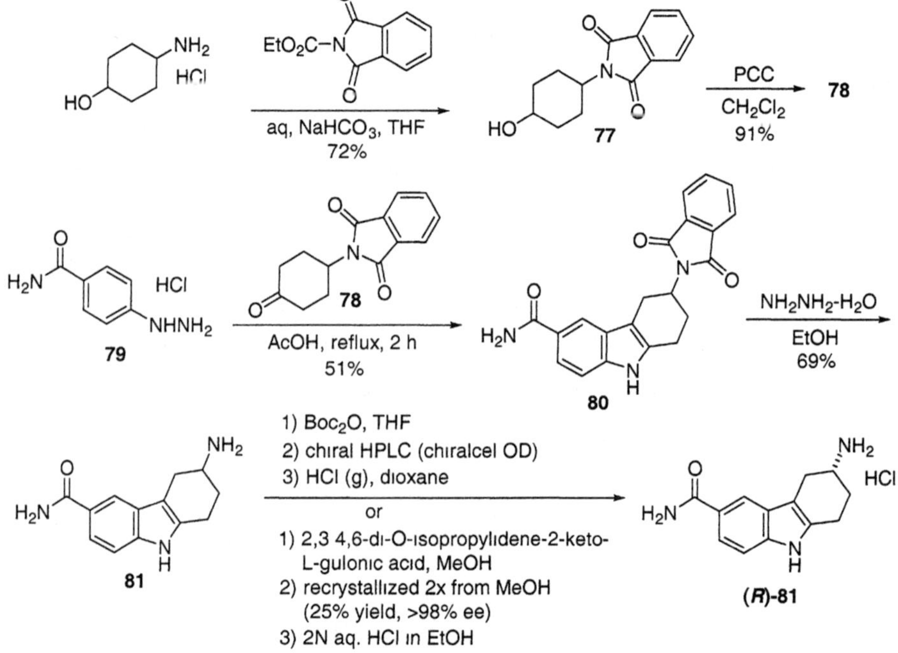

Scheme 26. Synthesis of amine intermediate **81** utilizing 4-phthalimido cyclohexanone (**78**) in the Fischer indole cyclization.

Reductive amination of benzaldehyde with (*R*)-**81** provided the corresponding benzyl amine, which was reductively alkylated with formaldehyde to give **82** (Scheme 27).[33] Compound **82** was debenzylated by hydrogenation in the presence of Pearlman's catalyst to afford frovatriptan ((*R*)-**6**). Alternatively, monomethylation of amine (*R*)-**81** was affected by treatment with carbon disulfide and dicyclohexylcarbodiimide in pyridine to provide isothiocyanate **83**, which was reduced with sodium borohydride to give (*R*)-**6**.

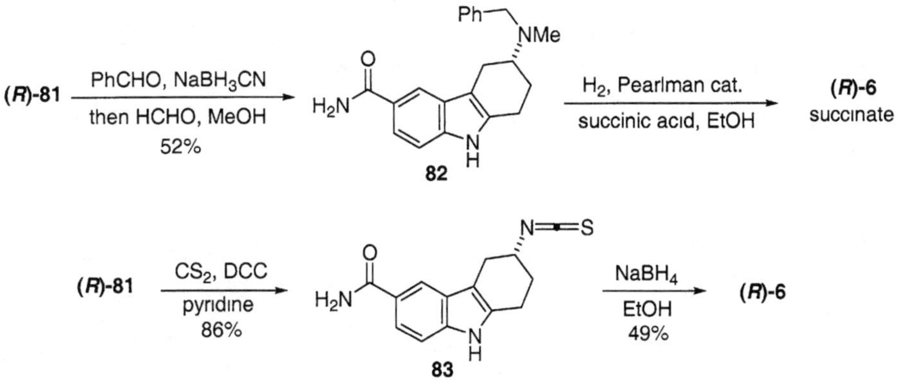

Scheme 27. Synthesis of optically pure frovatriptan ((R)-**6**) via methylation of intermediate **81**.

§12.8 Synthesis of eletriptan (7)[35–38]

Eletriptan (**7**, UK-116044) is a conformationally restricted analog of sumatriptan (**1**) and was discovered at Pfizer in the United Kingdom.[35] 5-Bromoindole (**37**) was acylated at the 3-position by reacting the magnesium salt of **37** with *N*-benzyloxycarbonyl-*D*-proline acid chloride (**84**) to form **85** in 50% yield. A process improvement for this step was recently published in the patent literature.[36] In this procedure, ethylmagnesium bromide (1M in methyl *tert*-butyl ether) and a solution of **84** in CH_2Cl_2 were added simultaneously to a solution of **37** in CH_2Cl_2 over 2 to 3 hours on opposite sides of the reaction vessel while maintaining the reaction temperature between 10–15 °C. This reaction was conducted on a multi-kilogram scale and crystalline **85** was isolated in 82% yield. The carbonyls in **85** were exhaustively reduced with $LiAlH_4$ to give the *N*-methylpyrrolidinylmethyl indole **86**. Cross-coupling of bromoindole **86** with phenylvinyl sulfone was accomplished under Heck conditions to produce **87**. Finally, the vinyl sulfone moiety was hydrogenated to afford eletriptan (**7**).

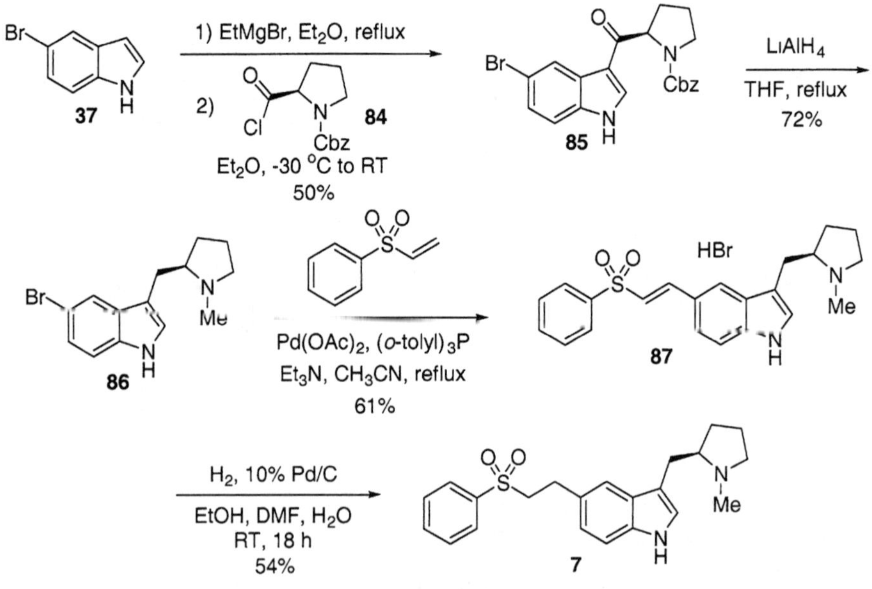

Scheme 28. Pfizer synthesis of eletriptan (**7**).

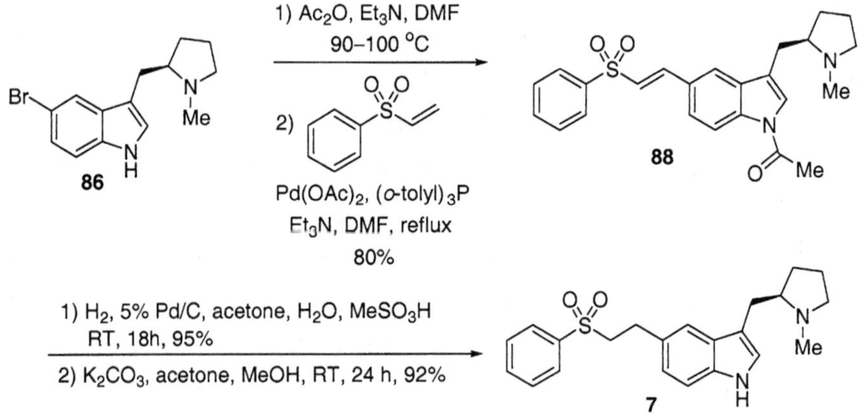

Scheme 29. Improved synthesis of eletriptan (**7**) using a Heck strategy.

An improved process for the preparation of eletriptan (**7**) is shown in Scheme 29.[35,37] A problem with the synthesis in Scheme 28 was that intermediate **87** was prone to dimerization, where the indole nitrogen of **87** added to the vinyl sulfone of another molecule. The formation of this impurity reduced the yield and complicated the purification. Therefore, indole **86** was protected with an acetyl group and then subjected to the Heck coupling with phenylvinyl sulfone to give **88**, which was no longer prone to dimerization. The reactive double bond in **88** was reduced by catalytic hydrogenation and the acetyl group was removed by hydrolysis to give eletriptan (**7**) of high purity in good yield.

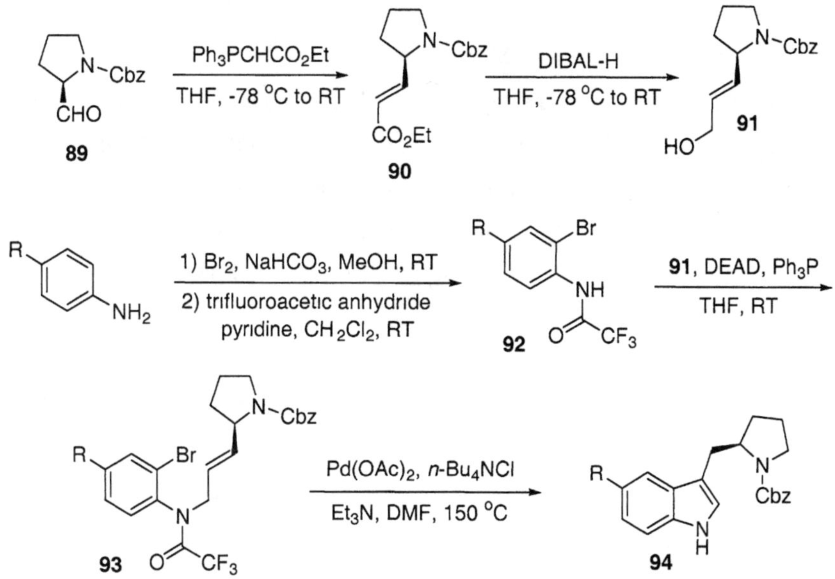

Scheme 30. Synthesis of the triptan core **94** using an intramolecular Heck reaction.

An alternative strategy for the preparation of the triptan core of eletriptan is shown in Scheme 30.[35] Wittig reaction of **89** with (carbethoxymethylene)triphenylphosphorane was initially conducted at −78 °C and warmed to RT to minimize possible racemization. The resultant α,β-unsaturated ester **90** was reduced with diisobutylaluminum hydride to give allylic alcohol **91**. The coupling partner **92** was prepared by bromination of an appropriately substituted aniline followed by reaction with trifluoroacetic anhydride. The trifluoroacetyl amide **92** was then coupled with alcohol **91** under Mitsunobu conditions to deliver the cyclization substrate **93**, which underwent an intramolecular Heck cyclization to furnish triptan **94**.

§12.9 References

1.   Diamond, S.; Wenzel, R. *CNS Drugs*, **2002**, *16*, 385–403.

2.   Longmore, J.; Dowson, A. J.; Hill, R. G. *Curr. Opin. CPNS Invest. Drugs*, **1999**, *1*, 39–53.

3.   Tepper, S. J.; Rapoport, A. M.; Sheftell, F. D. *Arch. Neurol.* **2002**, *59*, 1084–1088.

4.   Isaac, M.; Slassi, A. *IDrugs*, **2001**, *4*, 189–196.

5.   Milson, D. S.; Tepper, S. J.; Rapoport, A. M. *Exp. Opin. Pharmacother.* **2000**, *1*, 391–404. Tepper, S. J.; Rapoport, A. M. *CNS Drugs*, **1999**, *12*, 403–417. Deleu, D.; Hanssens, Y. *J. Clin. Pharmacol.* **2000**, *40*, 687–700.

6.   Yevich, J. P.; Yocca, F. D. *Curr. Med. Chem.* **1997**, *4*, 295–312.

7.   (sumatriptan) Dowles, M. D.; Coates, I. H. US 4816470 (**1989**).

8.   Oxford, A. W. US 5037845 (**1991**).

9.   Holman, N. J.; Friend, C. L. WO 01/34561.

10.   Pete, B.; Bitter, I.; Szantay, C., Jr.; Schon, I.; Toke, L. *Heterocycles*, **1998**, *48*, 1139–1149.

11.   Pete, B.; Bitter, I.; Harsanyi, K.; Toke, L. *Heterocycles*, **2000**, *53*, 665–673.

12.   *Drugs of the Future*, **1989**, *14*, 35–39.

13.   (zolmitriptan) Glen, R. C.; Martin, G. R.; Hill, A. P.; Hyde, R. M.; Woollard, P. M.; Salmon, J. A.; Buckingham, J.; Robertson, A. D. *J. Med. Chem.* **1995**, *38*, 3566–3580.

14.   Robertson, A. D.; Hill, A. P.; Glen, R. C.; Martin, G. R. US 5466699 (**1995**), US 5863935 (**1999**).

15.   Patel, R. US 6084103 (**2000**), US 6160123 (**2000**).

16.   *Drugs of the Future*, **1997**, *22*, 260–269.

17.   (naratriptan) Oxford, A. W.; Butina, D.; Owen, M. R. US 4997841 (**1991**).

18.   Blatcher, P.; Carter, M.; Hornby, R.; Owen, M. R. WO 95/09166.

19.   *Drugs of the Future*, **1996**, *21*, 476–479.

20.   Waterhouse, I.; Cable, K. M.; Fellows, I.; Wipperman, M. D.; Sutherland, D. R. *J. Labelled Compds. Radiopharm.* **1996**, *38*, 1021–1030.

21. (rizatriptan) Street, L. J.; Baker, R.; Davey, W. B.; Guiblin, A. R.; Jelley, R. A.; Reeve, A. J.; Routledge, H.; Sternfeld, F.; Watt, A. P.; Beer, M. S.; Middlemiss, D. N.; Noble, A. J.; Stanton, J. A.; Scholey, K.; Hargreaves, R. J.; Sohal, B.; Graham, M. I.; Matassa, V. G. *J. Med. Chem.* **1995**, *38*, 1799–1810.

22. Baker, R.; Matassa, V. G.; Street, L. J. US 5298520 (**1994**).

23. Baker, R.; Pitt, K. G.; Matassa, V. G.; Storey, D. E.; Olive, C.; Street, L. J.; Guiblin, A. R. EP 573221 (**1998**).

24. Chen, C. Y.; Lieberman, D. R.; Larsen, R. D.; Reamer, R. A.; Verhoeven, T. R.; Reider, P. J. *Tetrahedron Lett.* **1994**, *35*, 6981–6984.

25. Chen, C. Y.; Larsen, R. D.; Verhoeven, T. R. WO 95/32197.

26. Chen, C. Y.; Larsen, R. D. WO 98/06725.

27. *Drugs of the Future*, **1995**, *20*, 676–679.

28. (almotriptan) Bosch, J.; Roca, T.; Armengol, M.; Fernandez-Forner, D. *Tetrahedron*, **2001**, *57*, 1041–1048.

29. Forner, D. F.; Duran, C. P.; Soto, J. P.; Noverola, A. V.; Mauri, J. M. US 5565447 (**1996**).

30. Gonzalez, A. *Synth. Commun.* **1991**, *21*, 669–674.

31. *Drugs of the Future*, **1999**, *24*, 367–374.

32. (frovatriptan) King, F. D.; Gaster, L. M.; Kaumann, A. J.; Young, R. C. US 5464864 (**1995**).

33. Borrett, G. T.; Kitteringham, J.; Porter, R. A.; Shipton, M. R.; Vimal, M.; Young, R. C. US 5962501 (**1999**).

34. *Drugs of the Future*, **1997**, *22*, 725–728.

35. (eletriptan) Macor, J. E.; Wythes, M. J. US 5545644 (**1996**).

36. Perkins, J. F. EP 1088817 (**2001**).

37. Ogilvie, R. J. WO 02/50063.

38. *Drugs of the Future*, **1997**, *22*, 221–224.

Chapter 13. PDE 5 Inhibitors for Erectile Dysfunction: Sildenafil (Viagra®), Vardenafil (Levitra®), and Tadalafil (Cialis®)

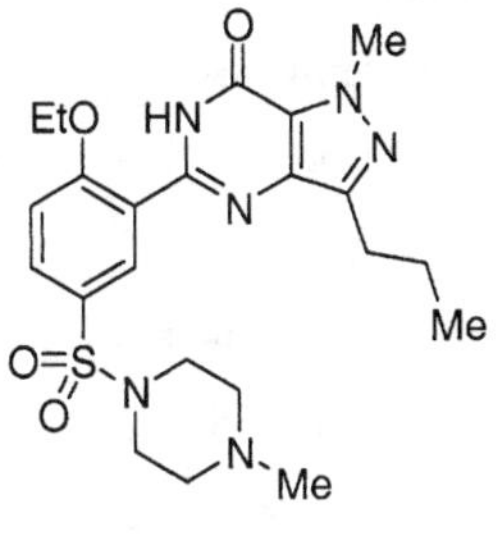

USAN Sildenafil Citrate
Trade Name: Viagra®
Company: Pfizer
Launched:1998
M.W. 474 58

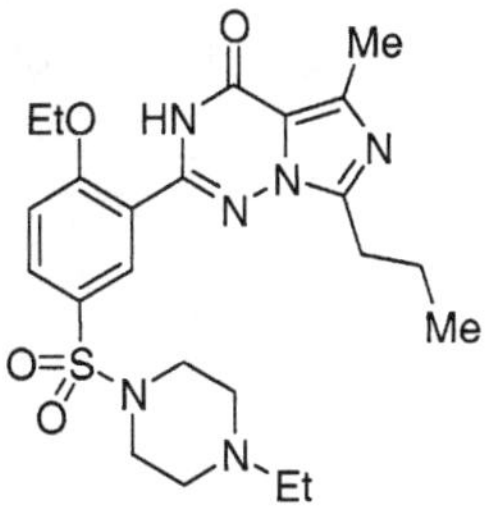

1

USAN: Vardenafil hydrochloride
Trade Name Levitra®
Company: Bayer/GlaxoSmithKline
Launched:2003
M.W. 488.60

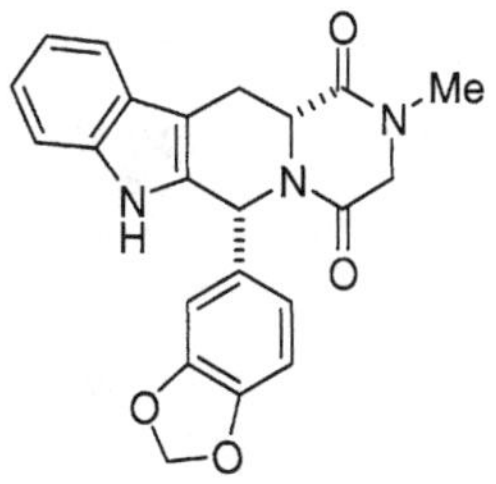

2

USAN· Tadalafil
Trade Name: Cialis®
Company: Icos/Eli Lilly
Launched:2003
M.W. 389.40

3

§13.1 Introduction[1–12]

Sildenafil (**1**) is a selective inhibitor of cyclic guanosine monophosphate (cGMP)-specific phosphodiesterase type 5 (PDE5). It was initially developed for the treatment of hypertension.[1,2] Levels of cGMP are controlled by phosphodiesterases, which convert cGMP into GMP (Figure 1). It was believed that a PDE inhibitor would prevent the breakdown of cGMP and the subsequent increase in cGMP concentration would allow smooth muscle cells in the kidney and blood vessels to relax, thus lowering blood pressure. Compounds were obtained that were quite selective for PDE5, however, it was shown that PDE5 is absent from the kidney so the focus of the project was switched to the treatment of angina by increasing blood flow to the muscles of the heart.[12] Initial clinical trials were disappointing, as the pharmacodynamic endpoints were not realized. However, some patients reported penile erections as a side-effect and the focus of the project changed to the treatment of male impotence.[1-3] At about this time, the role of nitric oxide in controlling erectile function was emerging. Sexual stimulation leads to the release of nitric oxide within the blood vessels of the penis, where it stimulates guanylate cyclase to increase cGMP levels in the corpus cavernosum.[2,4-10] This increase in cGMP causes smooth muscle relaxation, which increases blood flow to the penis and an erection results (Figure 1). It turns out there are high levels of PDE5 in the corpus cavernosum of the penis, which is able to degrade cGMP and cause termination of the erection. Nitric oxide production may be impaired in patients suffering from erectile dysfunction, leading to low levels of cGMP, which can be quickly degraded by PDE5. Inhibition of PDE5 by sildenafil (1) slows the breakdown of cGMP, allowing for higher concentrations to build up in the corpus cavernosum leading to an erection.

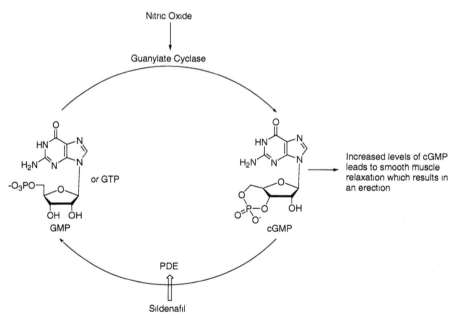

Figure 1.   Role of nitric oxide, cGMP, phosphodiesterase and sildenafil in erectile function.

Sildenafil (**1**)[4–6], the first effective oral treatment for erectile dysfunction, was introduced in 1998 and generated revenues of more than 1.7 billion dollars in 2002. Two more PDE5 inhibitors, vardenafil (**2**)[7,8] and tadalafil (**3**)[9], have recently entered the marketplace. They differ slightly in their potency, selectivity, pharmacokinetics, onset of action, duration of action and side-effect profile (Table 1). Vardenafil is the most potent at PDE5, but has the lowest oral bioavailability. Sildenafil and vardenafil are only about 10- and 15-fold selective against PDE6, respectively, but are very selective against PDE11 (> 700-fold). In contrast, tadalafil is very selective against PDE6 (> 100-fold), but is only about 10-fold selective against PDE11. Some of these differences can now be rationalized, since the crystal structures of **1–3** have been solved in the catalytic domain of human PDE5.[11] Sildenafil and vardenafil have a similar half-life of about 4 hours, however, tadalafil's half-life is much longer at 17.5 hours.

PDE5 inhibitor	PDE5 IC$_{50}$ (nM)	t$_{1/2}$ (h)	Oral bioavailability (%)	Approved doses (mg)
Sildenafil (**1**)	3.6–6.6	3–5 (3.8)	41	25, 50, 100
Vardenafil (**2**)	0.7	4–5 (4.7)	15	2.5, 5, 10, 20
Tadalafil (**3**)	2–6.7	17.5	NA	5, 10, 20

Table 1. Potency and pharmacokinetics of PDE5 inhibitors **1–3**.[4–10]

§13.2 Discovery and Synthesis of Sildenafil[2,4,12–16]

Sildenafil (**1**, UK-92480) was discovered at Pfizer in the United Kingdom. The starting point for the design of sildenafil was zaprinast, one of the few known inhibitors of PDE at the time (Table 2, entry 1).[2] Changing the core template to a pyrazolopyrimidinone provided a 10-fold boost in potency for PDE 5 with improved selectivity (entry 2). Increasing the length of the R$_1$ substituent from methyl to propyl provided another 10-fold enhancement in potency (entry 3). Many R$_2$ groups were tolerated but sulfonamides were chosen because they reduced the lipophilicity and increased the solubility of the compound. The *N*-methyl piperazine sulfonamide gave the best combination of potency (3.6 nM) and physical properties (log D 2.7) (entry 4). Sildenafil is only about 10 times more selective for PDE 5 than PDE 6 (from bovine retina), which may account for some of the visual side-effects that have been observed.

Zaprinast

entry	R$_1$	R$_2$	PDE 1[a]	PDE 2[a]	PDE 3[a]	PDE 4[a]	PDE 5[a]	PDE 6[a]	Log D
1	Zaprinast		9400	-	58,000	-	2000	-	-
2	Me	H	3300	-	>10,000	-	300	-	-
3	n-Pr	H	790	-	>10,000	-	27	-	4.0
4	n-Pr	SO$_2$R[b]	260	> 30,000	65,000	7300	3.6	38	27

[a]IC$_{50}$ in nM, [b]R = N-methyl piperazine

Table 2. SAR summary leading to the discovery of sildenafil (**1**).

The synthesis of sildenafil started with the reaction of 2-pentanone and diethyloxalate in the presence of sodium ethoxide to give diketoester **4** (Scheme 1).[12,13] Condensation of **4** with hydrazine afforded the pyrrazole **5**. Selective methylation of the 1-pyrrazole nitrogen with dimethylsulfate followed by hydrolysis of the ester gave **6**. The 4-position of the pyrrazole was nitrated with a mixture of fuming nitric acid and sulfuric acid and the amide was formed by treatment of the acid with thionyl chloride followed by ammonium hydroxide to furnish **7**. The nitro group was reduced with tin dichloride dihydrate in refluxing ethanol to give the requisite amine, which was acylated with 2-ethoxybenzoyl chloride using triethylamine and dimethylaminopyridine as base in dichloromethane. The resulting benzamido carboxamide **8** was cyclized to the pyrazolo[4,3-d]pyrimidin-7-one **9**, using aqueous sodium hydroxide and hydrogen peroxide in ethanol under reflux. Subsequently, **9** was selectively sulfonated para to the ethoxy group with chlorosulfonic acid to yield the sulfonyl chloride which was reacted with 1-methyl piperazine in ethanol to provide sildenafil (**1**).

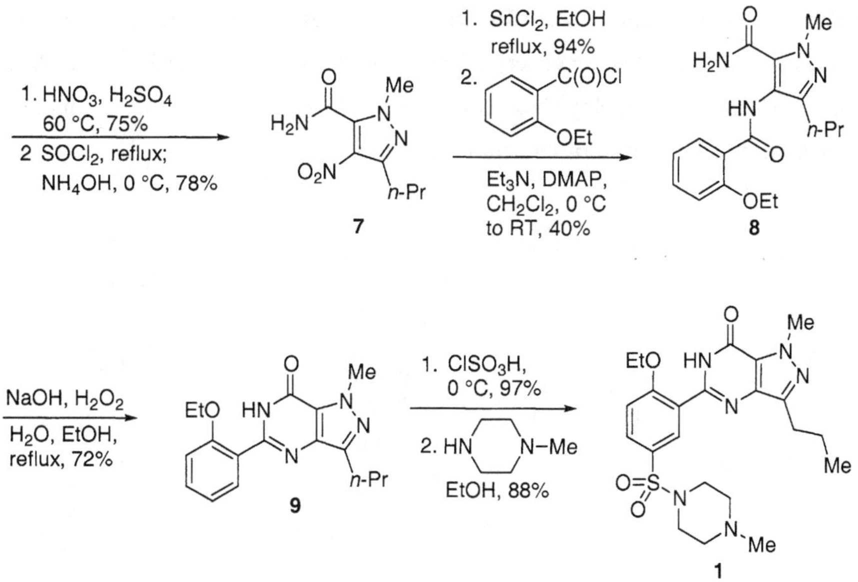

Scheme 1. Pfizer medicinal chemistry synthesis of sildenafil (1).

The medicinal chemistry route was not optimal for commercial manufacturing because the synthesis was completely linear and low yielding (11 steps, 4.2% overall yield from 2-pentanone). The Pfizer process group has developed a similar but more convergent and higher yielding synthesis of sildenafil (Scheme 2).[14] The chlorosulfonation was conducted early in the synthesis on 2-ethoxy benzoic acid thus avoiding potential toxic intermediates in the final step as in the initial discovery route. Addition of 1-methyl piperazine to the sulfonyl chloride was initially performed in the presence of triethylamine, however this method provided the hydrochloride-triethylamine double salt of **10**, which was very insoluble and was difficult to use in subsequent steps. This salt could be transformed to **10** by treatment with water. An improved method employed 1 equivalent of sodium hydroxide to give **10** directly. Hydrogenation of **7** and acylation of the resulting amine with the activated acid of **10** were performed in one step to deliver the amide **11** in 90% yield. All the reactions were conducted using ethyl acetate as solvent, which dramatically simplified the process and allowed for efficient solvent recovery. The final cyclization was conducted using potassium t-butoxide in refluxing t-butanol to provide sildenafil (**1**) of clinical quality directly from the reaction with no further purification. The commercial route was further optimized to minimize waste and maximize solvent recovery. Pfizer was awarded the Crystal Faraday Award for Green Chemical Technology by the Institute of Chemical Engineers in recognition of the environmentally-friendly manufacturing of sildenafil.

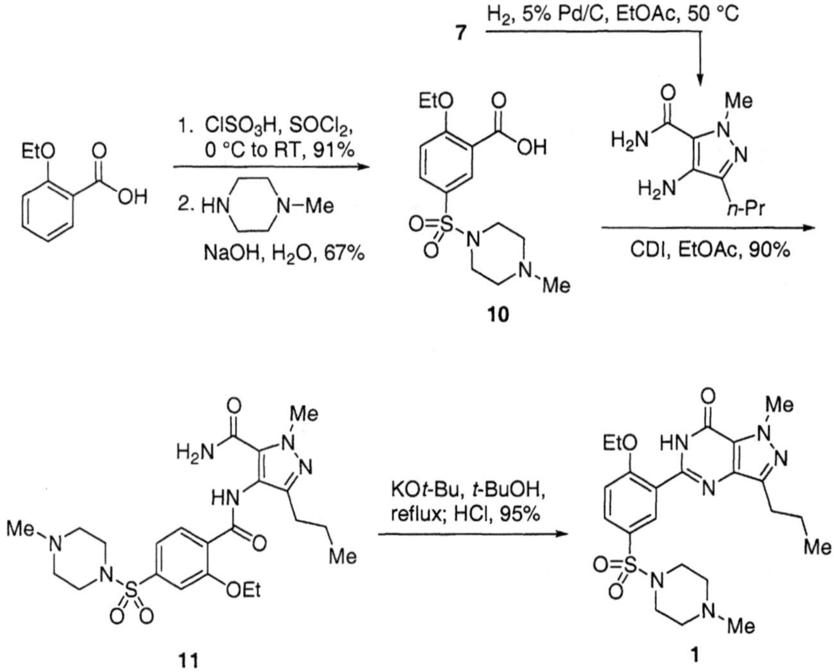

Scheme 2. Commercial synthesis of sildenafil (1).

§13.3 Synthesis of Vardenafil[17-20]

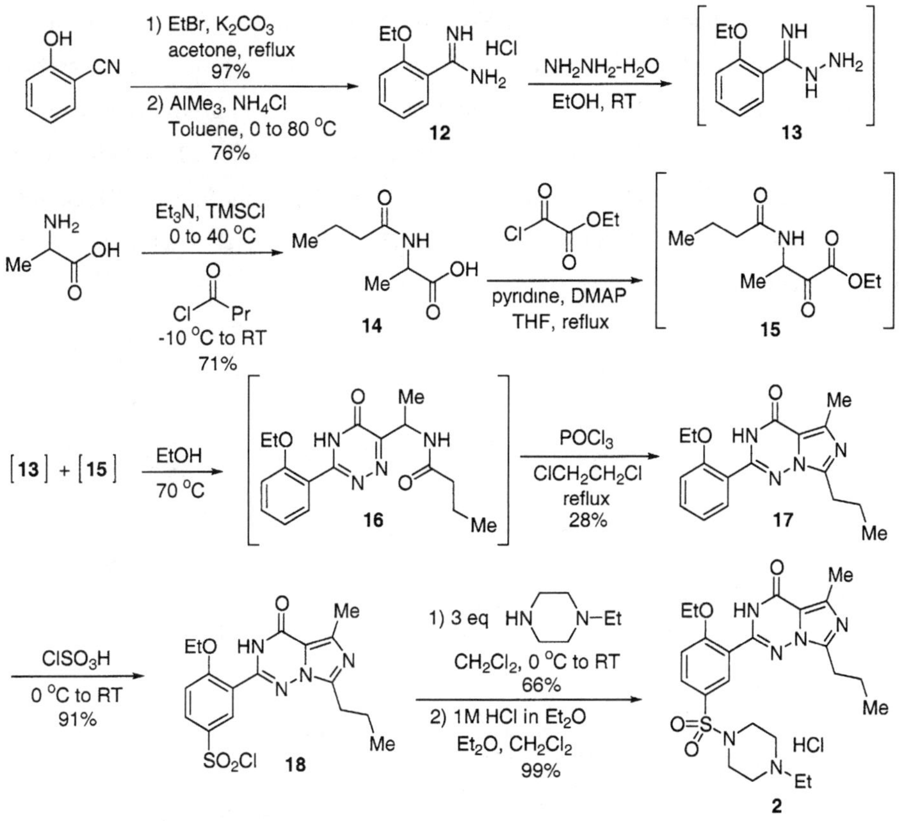

Scheme 3. Bayer synthesis of vardenafil (2).

Vardenafil (2, Bay-38-9456) was discovered at Bayer in Germany and is being co-marketed with GlaxoSmithKline. The structure of vardenafil (2) looks very similar to sildenafil (1), except 2 contains a slightly different purine-isosteric imidazo[5,1-ƒ][1,2,4]triazin-4(3H)-one heterocyclic core. The synthesis shown in Scheme 3[17,18] commenced with the alkylation of 2-hydroxybenzonitrile with ethyl bromide, followed by addition of ammonia to the nitrile functionality using AlMeClNH₂, generated *in situ* by reacting ammonium chloride with trimethylaluminium, to give amidine 12. Meanwhile, D,L-alanine was acylated with butyryl chloride to provide 14, which

underwent a Dakin-West reaction with ethyl oxalyl chloride to furnish the intermediate
α-oxoamino ester **15**.   A solution of crude **15** in ethanol was added to
carboximidohydrazide **13**, formed *in situ* by addition of hydrazine hydrate to amidine **12**
in ethanol, and the mixture was heated at 70 °C for 4 hours to give the condensation
product **16**. Intermediate **16** was cyclized to the imidazotriazinone **17** using phosphorous
oxychloride.   Compound **17** was formed in 28% yield from **12** and **14**, when
intermediates **13**, **15**, and **16** were taken to the next step without purification. Reaction of
17 with chlorosulfonic acid provided the sulfonyl chloride **18**, which was treated with *N*-
ethylpiperazine to furnish vardenafil (**2**).

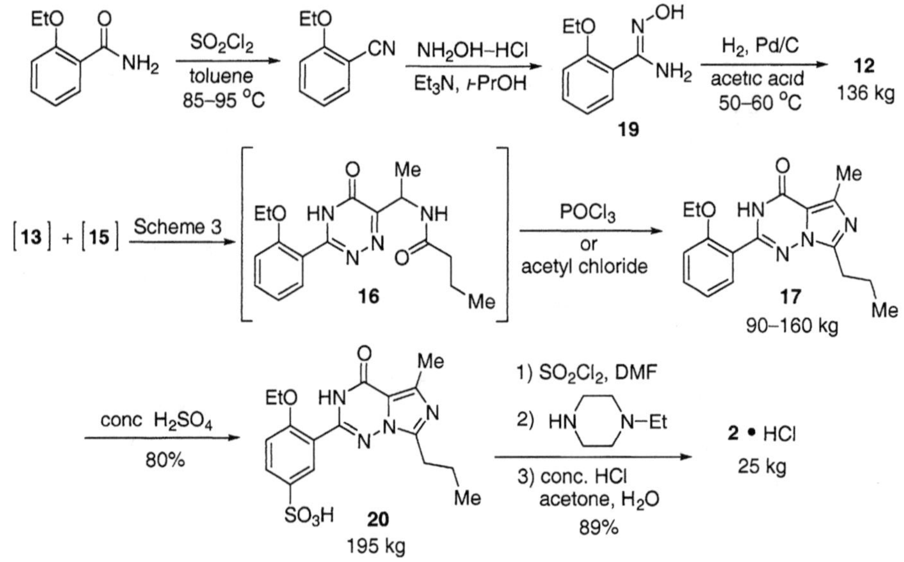

Scheme 4. Bayer large scale synthesis of vardenafil (**2**).

A modified synthesis of vardenafil (**2**) on a multi-kilogram scale has recently been
reported by a group from Bayer (Scheme 4).[19]   Benzamidine **12** was prepared in a
different manner than described in Scheme 3, because it is difficult to use
trimethylaluminium on a large scale.   Thus, 2-ethoxybenzamide was dehydrated with
thionyl chloride to give 2-ethoxybenzonitrile, which was treated with hydroxyl amine
hydrochloride to afford the *N*-hydroxybenzamidine **19**. Catalytic hydrogenation of **19**

provided benzamidine **12** on a 136 kilogram scale. Intermediate **15** was condensed with the imidohydrazide intermediate **13** in a similar manner as described in Scheme 3 to give **16**. Compound **16** was cyclized using phosphorous oxychloride or acetyl chloride to deliver **17** in quantities ranging from 90 to 160 kilograms. Compound **17** was sulfonated with H_2SO_4 to furnish the sulfonic acid **20** on a 195 kilogram scale. Treatment of **20** with thionyl chloride and DMF provided the sulfonyl chloride, which was reacted with *N*-ethylpiperazine in the same pot to give 26 kilograms of vardenafil (**2**) in 93% yield. The hydrochloride salt of **2** was prepared using concentrated HCl in acetone and water.

§13.4 Synthesis of Tadalafil[21–23]

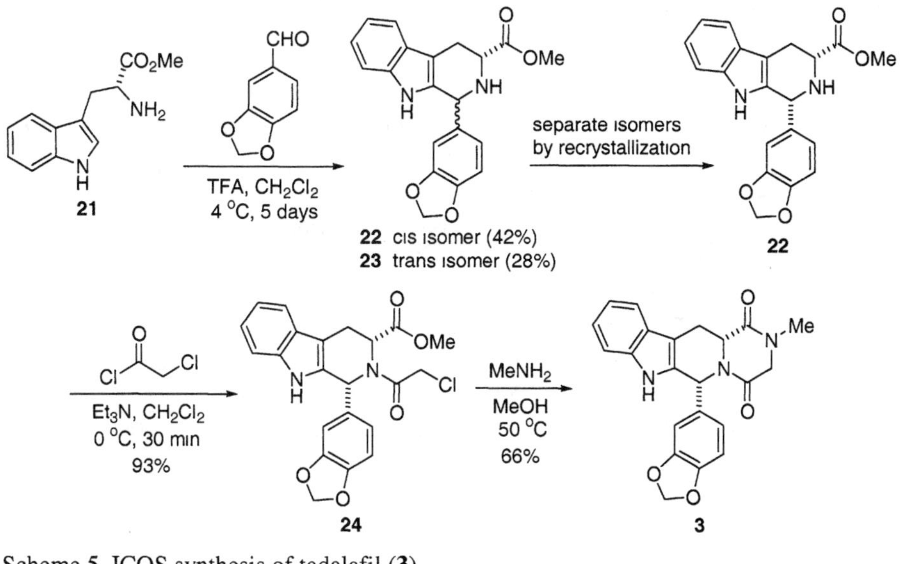

Scheme 5. ICOS synthesis of tadalafil (**3**).

Tadalafil (**3**, IC-351) was discovered at ICOS and is being developed and marketed as a joint venture with Eli Lilly. *D*-tryptophan methyl ester (**21**) was condensed with piperonal by means of trifluoroacetic acid and the C2 carbon of the indole adds to the resulting iminium ion to give a mixture of cis β-carboline **22** and trans isomer **23** (Scheme 5).[21] The desired cis isomer **22** can be isolated by chromatography or crystallization in 42% yield. Compound **22** was acylated with chloroacetyl chloride to provide **24**, which was cyclized with methylamine in methanol to deliver tadalafil (**3**). The undesired trans β-carboline **23** can be equilibrated to give the cis isomer **22** by treatment with aqueous HCl at 60 °C for 36 hours (Scheme 6).

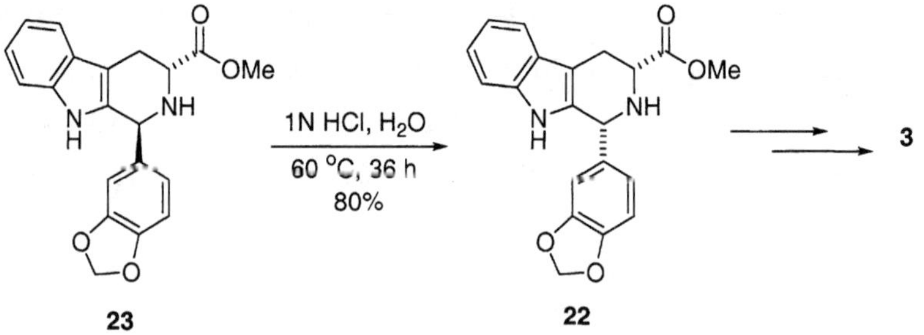

Scheme 6. Equilibration of trans isomer **23** to cis isomer **22**.

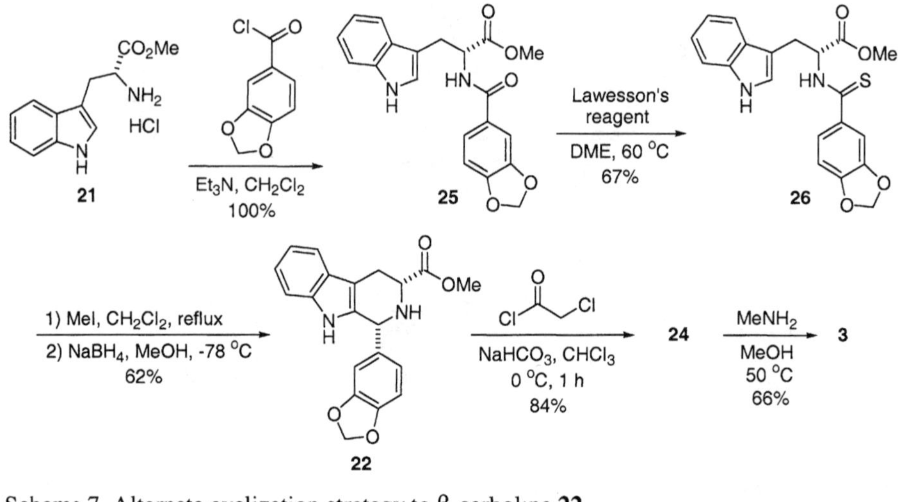

Scheme 7. Alternate cyclization strategy to β-carboline **22**.

An alternate step-wise cyclization strategy to form β-carboline **22** is shown in Scheme 7.[21] Acylation of *D*-tryptophan (**21**) with piperonyloyl chloride afforded amide **25**, which was converted to thioamide **26** with Lawesson's reagent. Thioamide **26** was treated with methyl iodide in refluxing CH₂Cl₂ to give a α-thiomethylimmonium ion, which was trapped by the intramolecular addition of indole. The thiomethyl group was then eliminated and the resulting cyclic imine intermediate was reduced with sodium borohydride predominately from the least hindered face to deliver the cis carboline **22**. Compound **22** was then acylated with chloroacetyl chloride and cyclized with methyamine as in Scheme 5 to give tadalafil (**3**).

§13.5 References

1.   Kling, J. *Modern Drug Discovery*, **1998**, *1*, 31–38.

2.   Campbell, S. F. *Clinical Science*, **2000**, *99*, 255–260.

3.   Ellis, P.; Terrett, N. K. WO 94/28902.

4.   (sildenafil) Cartledge, J.; Eardley, I. *Exp. Opin. Pharmacother.* **1999**, *1*, 137–147.

5.   Noss, M. B.; Christ, G. J.; Melman, A. *Drugs Today*, **1999**, *35*, 211–217.

6.   Brock, G. *Drugs Today*, **2000**, *36*, 125–134.

7.   (vardenafil) Martin-Morales, A.; Rosen, R. C. *Drugs Today*, **2003**, *39*, 51–59.

8.   Young, Y. M. *Expert. Opin. Investig. Drugs*, **2002**, *11*, 1487–1496.

9.   (tadalafil) Pomerol, J. M.; Rabasseda, X. *Drugs Today*, **2003**, *39*, 103–113.

10.   Rotella, D. P. *Nature Rev. Drug Discov.* **2002**, *1*, 674-682.

11.   Sung, B.-J.; Hwang, K. Y.; Jeon, Y. H.; Lee, J.; *et. al. Nature*, **2003**, *425*, 98-102.

12.   (sildenafil) Bell, A. S.; Brown, D.; Terrett, N. K. EP 463756 (**1995**), US 5250534 (**1993**).

13.   Terrett, N. K.; Bell, A. S.; Brown, D.; Ellis, P. *Bioorg. Med. Chem. Lett.* **1996**, *6*, 1819–1824.

14.   Dale, D. J.; Dunn, P. J.; Golightly, C.; Hughes, M. L.; Levett, P. C.; Pearce, A. K.; Searle, P. M.; Ward, G.; Wood, A. S. *Org. Process Res. Dev.* **2000**, *4*, 17-22. Dunn, P. J.; Wood, A. S. EP 812845 (**1999**).

15.   Baxendale, I. R.; Ley, S. V. *Bioorg. Med. Chem. Lett.* **2000**, *10*, 1983-1986.

16.   *Drugs of the Future*, **1997**, *22*, 138–143.

17.   (vardenafil) Niewohner, U.; Es-Sayed, M; Haning, H.; Schenke, T.; Schlemmer, K.-H.; Keldenich, J.; Bischoff, E.; Perzborn, E.; Dembowsky, K.; Serno, P.; Nowakowski, M. US 6362178 (**2002**).

18.   Haning, H.; Niewohner, U.; Schenke, T.; Es-Sayed, M; Schmidt, G.; Lampe, T.;
       Bischoff, E. *Bioorg. Med. Chem. Lett.* **2002**, *12*, 865–868.

19.   Nowakowski, M.; Gehring, R.; Heilmann, W.; Wahl, K.-H. WO 02/50076.

20.   *Drugs of the Future*, **2001**, *26*, 141–144.

21.   (tadalafil) Daugan, A. C.-M. US 5859006 (**1999**); US 6025494 (**2000**).

22.   Maw, G. N.; Allerton, C. M. N.; Gbekor, E.; Million, W. A. *Bioorg. Med. Chem.
       Lett.* **2003**, *13*, 1425–1428.

23.   *Drugs of the Future*, **2001**, *26*, 15–19.

Chapter 14. Antiasthmatics

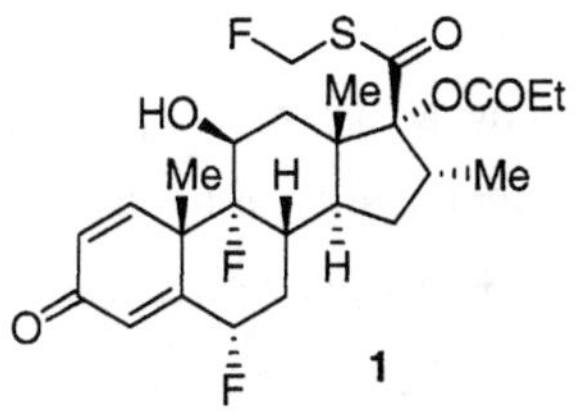

USAN: Fluticasone propionate
Trade Name· Flovent® or Flonase®
GlaxoSmithKline
U.S. Approval: 1994
M W. 500.57

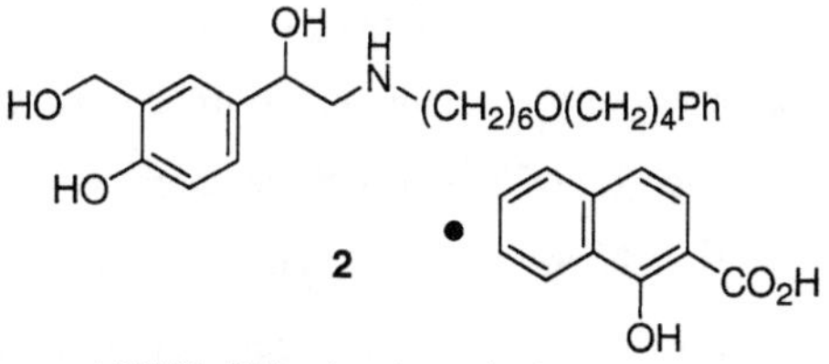

USAN· Salmeterol xinafoate
Trade Name: Serevent®
GlaxoSmithKline
U S. Approval· 1994
M.W. 415.57 (parent)

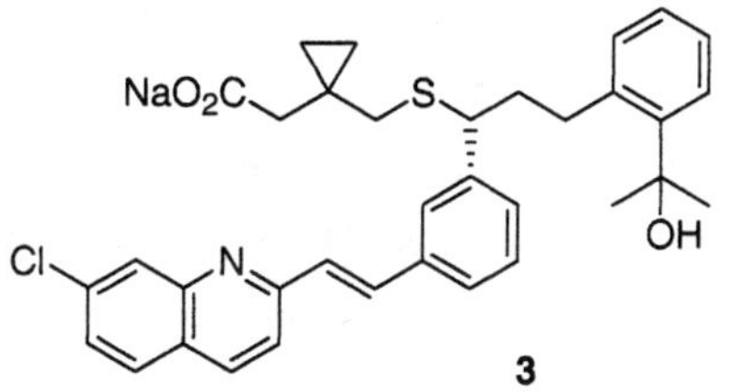

USAN· Montelukast sodium
Trade Name: Singulair®
Merck
U.S. Approval: 1998
M.W. 608.17

§14.1 Introduction

Asthma is a chronic inflammatory condition characterized by bronchial hyper-responsiveness and reversible airway obstruction. Cytokine release from a variety of cell types such as eosinophils, lymphocytes and other inflammatory cells produces epithelial sloughing, plasma protein extravasation from the tracheobronchial microcirculation and airway remodeling.[1,2] Bronchial mucosal inflammation is present in all patients. The primary goal of asthma management is to maintain control of the disease process by reducing symptoms and improving lung function.

The pharmacological control of asthma can be achieved in most patients with chromones and inhaled glucocorticoids. Corticosteroids are considered the mainstay of asthma therapy. The introduction of inhaled preparations made this class of drugs the most suitable for the treatment of asthmatic patients. They alleviate the major symptoms of the disease by reducing airway reactivity while restoring the integrity of the airways. The mechanism of action of these agents is not well understood. The clinical efficacy of these agents is probably the result of their inhibitory effect on leukocyte recruitment into the airways. However, therapeutic doses of oral glucocorticoids are associated with a wide range of adverse effects such as Cushing's syndrome, altered lipid and bone metabolism, bone erosions and vascular effects.[3]   Glucocorticoids such as beclomethasone dipropionate have significant bioavailability and when one considers the surface area of the tracheobronchial mucosa, significant plasma levels and systemic side-effects occur at therapeutic doses. Plasma levels of cortisone are reduced and the hypothalamic-pituitary-adrenal axis is suppressed. However, the onset of osteoporosis and reduced bone growth in children is by far the most serious adverse event. The solution to the high bioavailability of these agents was the development of fluticasone propionate (1), a prodrug that results in much lower oral bioavailability.[3]

The β2-adrenergic agonists are the most prescribed class of drugs for the treatment of asthma. They are preferred both for the rapid relief of symptoms and for the level of bronchodilation achieved in patients with bronchial asthma. These drugs produce their effects through stimulation of specific β2-adrenergic receptors located in the plasma membrane, resulting in alterations in adenylyl cyclase and elevations in intracellular AMP. Cyclic AMP is responsible for the relaxation of smooth muscle with bronchodilation in the bronchi and a reduction in mucus viscosity. Long-acting β2-adrenergic agonists, such as salmeterol xinafoate (2), are very lipophilic and have a high affinity for the receptor by a different mechanism. However, these treatments also suffer from a variety of side-effects. The widespread distribution of β2-adrenergic receptors results in a number of undesired responses when these agents are absorbed into the systemic circulation. Tremor is the most common side-effect and results from stimulation of the β2-receptor in skeletal muscle. The most serious side-effects are cardiac in nature (increased heart rate, tachyarrhythmias) that result from stimulation of the β2-receptor in the heart. Most of these side-effects disappear with long-term use and do not have any long-term health consequences.[3]

The newest therapy available for the treatment of asthma arises from the recognition of the role of the leukotrienes (LTs) in the initiation and propagation of airway inflammation. The evidence to support the role of leukotrienes in bronchial asthma includes: a) cells known to be involved in asthma produce LTs; b) cysteinyl LTs (LTC$_4$, LTD$_4$ and LTE$_4$) cause airway abnormalities and mimic those seen in asthma; c) the production of LTs is increased in the airways of people with asthma. The leukotrienes exert their effects through G protein coupled receptors (GPCRs) regulating a signal transduction pathway that ultimately causes calcium release from the cells. There are two classes of leukotriene receptors, BLT$_1$ receptors and CysLT receptors 1 and 2. It is these latter receptors that mediate the actions of the cysteinyl leukotrienes in asthma. The first generation of CysLT receptor antagonists were neither potent nor selective. The second generation of antagonists is approximately 200 times more potent than the first generation. Montelukast sodium (3) is a prime example of this class and is typified by high intrinsic potency, good oral bioavailability and long duration of action.[4]

§14.2 Synthesis of fluticasone propionate (1)

The first inhaled glucocorticoid, beclomethasone dipropionate, revolutionized asthma therapy, when it was found that topical delivery to the lung resulted in reduced systemic side-effects (adrenal suppression, osteoporosis and growth inhibition) typically seen with oral steroid treatments. Interestingly, a further reduction in systemic exposure was achieved with the introduction of fluticasone propionate (1). The evolution of this drug stemmed from observations with the steroid 17-carboxylates that showed that these esters were active topically when esterified, while the parent acids were inactive. Thus it was realized that enzymatic hydrolysis of the ester would lead to systemic deactivation. SAR studies led to a series of carbothioates, which were very active in vivo when topically applied to rodents, but were inactive after oral administration. It was shown that fluticasone propionate (1) underwent first pass metabolism in the liver to the corresponding inactive 17β-carboxylic acid (1a) (Scheme 1). This observation was confirmed by experiments that showed it was rapidly metabolized by mouse, rat or dog liver homogenates.[5]

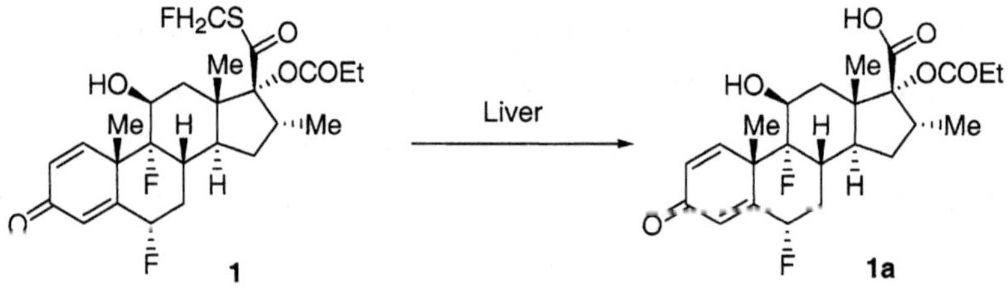

Scheme 1. Metabolic deactivation of fluticasone propionate (1) in the liver.

The synthesis of fluticasone propionate (1) utilizes commercially available flumethasone (4) (Scheme 2). Oxidation of 4 with periodic acid gave the etianic acid 5[6], whose imidazolide when treated with hydrogen sulfide gas gave the carbothioic acid 6. Treatment with excess propionyl chloride followed by aminolysis of the mixed anhydride with diethylamine gave 7. Alkylation with bromochloromethane gave the chloromethyl carbothioate 8, which was converted to the iodomethyl ester 9 by treatment with sodium iodide. Conversion to fluticasone (1) was accomplished by treatment with silver fluoride in acetonitrile. Alternatively, fluticasone (1) was also prepared directly from the potassium salt of carbothioic acid 7 using bromofluoromethane.[7]

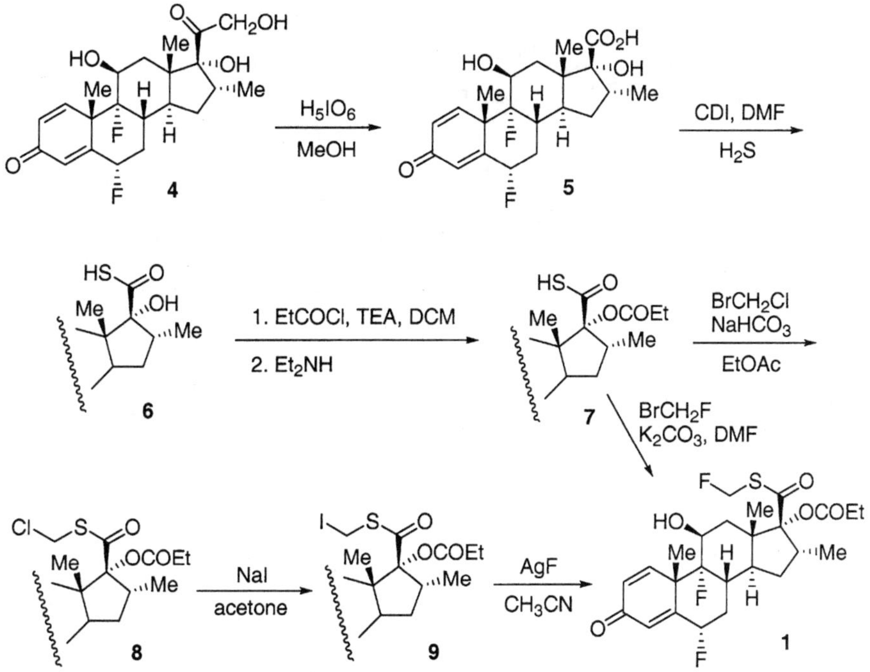

Scheme 2. Synthesis of fluticasone propionate (1) from flumethasone (4).

§14.3 Synthesis of salmeterol xinafoate (2)

Salmeterol xinafoate (2) is marketed as a racemic drug under the trade name, Serevent® by GlaxoSmithKline. Furthermore, the combination of salmeterol xinafoate (2) and fluticasone propionate (1) was approved in the U.S. in 2000 and is marketed as Advair®. The original racemic synthesis of salmeterol appeared in the patent literature and started with alkylation of 4-phenyl-1-butanol (10) with an excess of 1,6-dibromohexane in the presence of sodium hydride to give the bromoether 11 (Scheme 3).[8,9] Alkylation of the phenethanolamine 12 with bromoether 11 using potassium iodide and triethylamine in DMF provided salmeterol (2). Treatment of 2 with 1-hydroxy-2-naphthoic acid in methanol gave the corresponding xinafoate salt. The synthesis of 12 was accomplished starting from the acyl salicylic acid 13. Bromination gave the acyl bromide 14, which was reacted with benzylamine followed by LiAlH₄ reduction to yield the corresponding alcohol 15. Finally, debenzylation of 15 provided 12.[10]

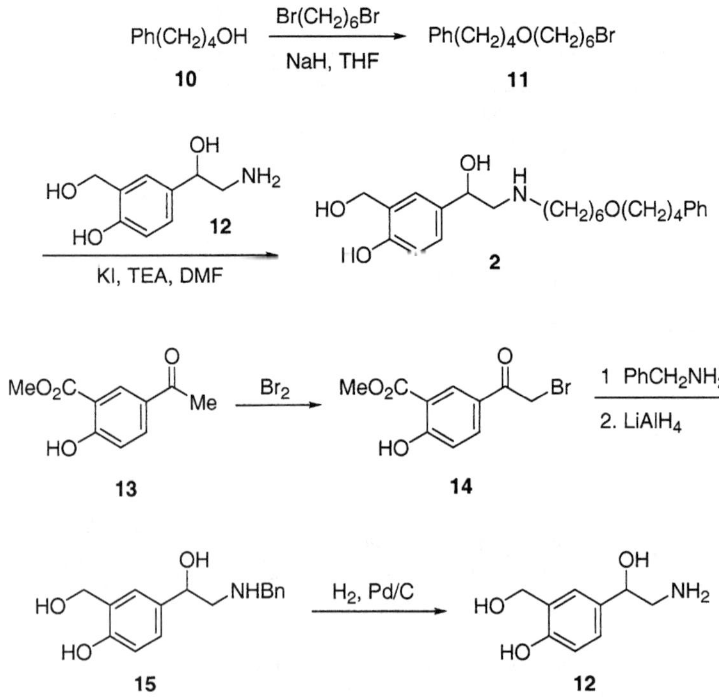

Scheme 3. Original synthesis of salmeterol xinafoate (**2**).

Ruoho and Rong have described a shorter route to salmeterol (Scheme 4).[11] Friedel-Crafts acylation of salicylaldehyde (**16**) with bromoacetyl bromide in the presence of aluminum chloride gave the acetophenone **17**. Alkylation of amine **18** with bromoacetyl **17** in refluxing acetonitrile gave the ketone **19**. Reduction of **19** with sodium borohydride in methanol followed by catalytic hydrogenolysis of the benzyl group over 10% Pd/C gave salmeterol (**2**).

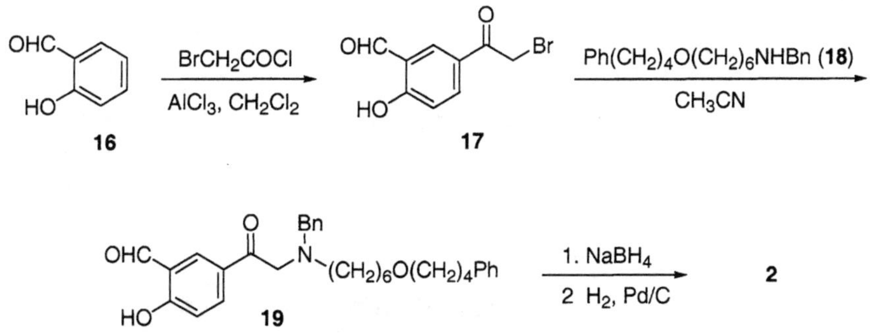

Scheme 4. Synthesis of salmeterol (2) from salicylaldehyde.

There has been considerable interest in the enantiomers of salmeterol (2). Salmeterol (2) is a long acting, potent β2 adrenoreceptor agonist. The (R)-enantiomer is more potent than the (S)-enantiomer, however, it has been recently claimed that the (S)-enantiomer has a higher selectivity for β2 receptors and thus exhibited fewer side-effects than the (R)-enantiomer or the racemate.[12]

The original racemic patents described the use of resolution to give a chiral oxirane, such as 25, as an intermediate or the use of a chiral auxiliary (20) to produce the salmeterol enantiomers. Alkylation of chiral amine 20 with 2-benzyloxy-5-(2-bromo-acetyl)-benzoic acid methyl ester, followed by diastereoselective reduction of the ketone with lithium borohydride furnished intermediate 21 after chromatographic separation of the diasteromers. Removal of the benzyl group and the chiral auxiliary was accomplished by catalytic hydrogenolysis to give (R)-salmeterol (2) as shown in Scheme 5.[8, 9]

Procopiou et al. have also employed a chiral auxiliary mediated diastereoselective reduction using NaBH$_4$-CaCl$_2$ as the reducing agent.[13] This method was subsequently applied to an enantioselective synthesis of (R)-salmeterol (2) using a sequence of supported agents and scavenging agents and employing CaCl$_2$ and a borohydride exchange resin (BER).[14]

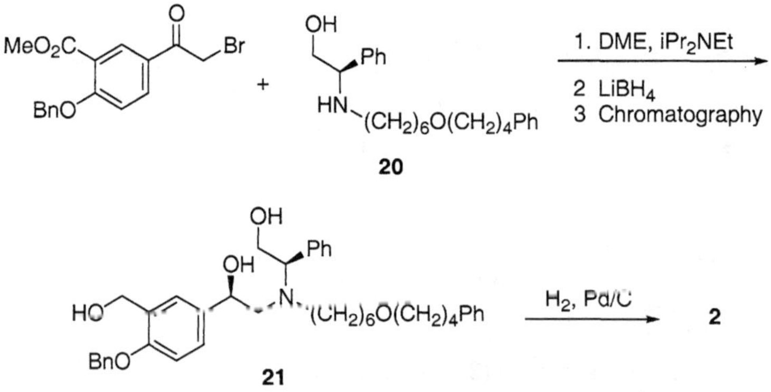

Scheme 5. Synthesis of the salmeterol enantiomers using a chiral auxiliary.

Helquist *et al.* described the first enantioselective synthesis[15] of either enantiomer using an asymmetric borane reduction catalyzed by the chiral oxazaborolidine catalysts developed by Corey (Scheme 6).[16] Treatment of the bromoketone **22** with borane and a catalytic amount of (*R*)-oxazaborolidine catalyst (**23**) gave the bromohydrin **24** in quantitative yield and > 94% *ee*. The (*S*)-catalyst can be employed to give (*S*)-salmeterol (**2**). Treatment of the bromohydrin **24** with a sodium hydride suspension in THF gave the epoxide **25** with an optical purity > 95%. This was then treated with the amine **18** in a regioselective ring opening to give the alcohol **26** in > 94% *ee*. Reduction of the ester with LiAlH₄ and hydrogenolysis of the benzyl protecting groups gave (*R*)-salmeterol (**2**).

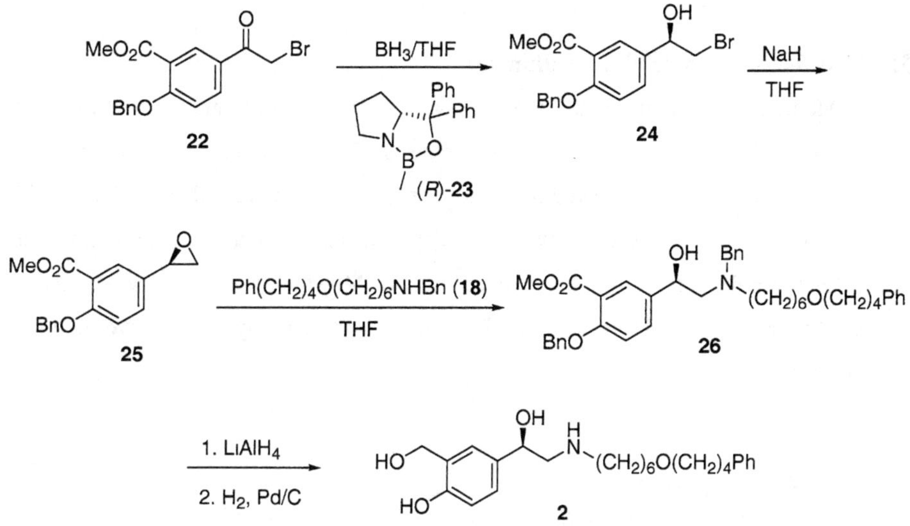

Scheme 6. Enantioselective synthesis of (R)-salmeterol (2).

Biotransformation pathways have also been used to establish this chiral center (Scheme 7). The bromoacetophenone 27 was mixed with sodium laurel sulfate and added to a microbial culture of *Rhodotorula rubra* to produce enantiomerically pure alcohol 28 with 95% *ee* that was eventually converted to (R)-salmeterol (2).[17] In a similar transformation, the azidoketone 29 was enantioselectively reduced, using the microorganism *Pichia angusta*, to the alcohol 30 with > 98% *ee* that was eventually converted to (S)-salmeterol (2).[18]

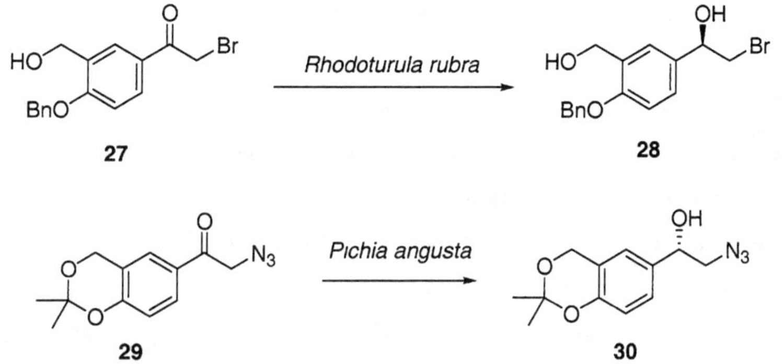

Scheme 7. Enzymatic enantioselective reduction to give salmeterol intermediates.

§14.4 Synthesis of montelukast sodium (3)

Montelukast sodium (**3**) has the (*R*) absolute configuration and its synthesis was first described by Labelle *et al.*[19,20] The synthesis of the side-chain commenced with reduction of the diester **31** followed by monobenzoylation of the resulting diol to give **32** (Scheme 8). Alcohol **32** was mesylated and then displaced with cyanide to give the nitrile **33**. Cyanide hydrolysis and esterification gave **34**, which was again mesylated and displaced with AcSCs in DMF to give **35**. Deprotection using hydrazine gave the thiol side-chain **36**.

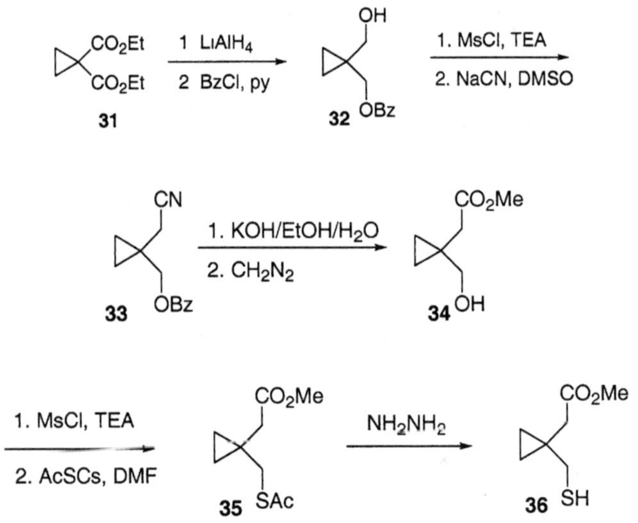

Scheme 8. Synthesis of the montelukast side-chain **36**.

The synthesis of the montelukast core and coupling of the thiol side-chain **36** to complete the synthesis of **3** is shown in Scheme 9.[19,21] Quinoline **37** was reacted with isophthalaldehyde in refluxing acetic anhydride to provide **38**.[22] Vinylmagnesium bromide was added to aldehyde **38** to give the allylic alcohol **39**, which underwent a palladium catalyzed cross-coupling with methyl 2-bromobenzoate in the presence of lithium chloride to give **40**. The ketoester **40** was enantioselectively reduced using the Corey reduction followed by addition of methyl magnesium bromide to the ester to afford the key chiral alcohol **41**.[20]   The secondary alcohol was protected as the *t*-butyldimethylsilyl (TBS) ether and the tertiary alcohol as the THP ether to give **42**. The TBS protecting group was removed and the alcohol was converted to the mesylate **43**. Thiol **36** was added to mesylate **43** in the presence of cesium carbonate to give the advanced ester intermediate **44**. Deprotection, ester hydrolysis and salt formation gave montelukast sodium (**3**).

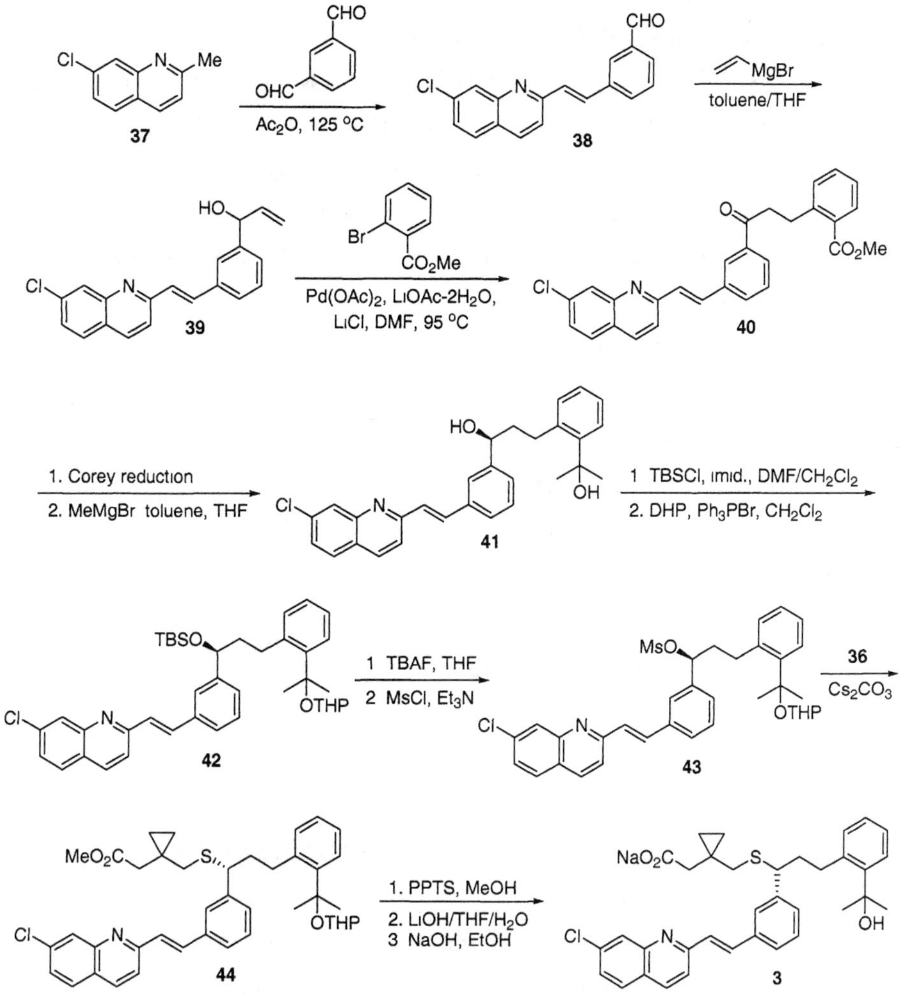

Scheme 9. Synthesis of montelukast (**3**).

§14.5 References

1.   Holgate, S. *Thorax,* **1993**, *48*, 103–109.

2.   Bousquet, J.; Chanez, P.; Lacoste, J. Y.; *et al. Allergy,* **1992**, *47*, 3–11.

3.   Oliveira, E.; Martin, R. J.; Martin, V. *Drugs of Today,* **1998**, *34*, 203–223.

4.   Balkan, A.; Berk, B. *Curr. Med. Chem.* **2003**, *2*, 9–18.

5   Phillipps, G. H. *Am. Rev. Respir. Med.* **1990**, *84 (Suppl. A)*, 19–23.

6.   Kertesz, D. J.; Marx, M. *J. Org. Chem.* **1986**, *51*, 2315–2328.

7.   Phillipps, G. H.; Bailey, E. J.; Bain, B. M.; Borella, R. A.; Buckton, J. B.; Clark, J. C.; Doherty, A. E.; English, A. F.; Fazakerley, H.; Laing, S. B.; Lane-Allman, E.; Robinson, J. D.; Sandford, P. E.; Sharratt, P. J.; Steeples, I. P.; Stonehouse, R. D.; Williamson, C. W. *J. Med. Chem.* **1994**, *37*, 3717–3729.

8.   Skidmore, I. F.; Lunts, L. H. C.; Finch, H.; Naylor, A. US 4992474 (**1991**).

9.   Evans, B. EP 0422889 A2 (**1990**).

10.   Collin, D. T.; Hartley, D.; Jack, D.; Lunts, L. H. C.; Press, J. C.; Ritchie, A. C.; Toon, P. *J. Med. Chem.* **1970**, *13*, 674–680.

11.   Rong, Y.; Ruoho, A. E. *Synth. Commun.* **1999**, *29*, 2155–2162.

12.   Jerusi, T. P. WO 99/13867 (**1999**).

13.   Bream, R. N.; Ley, S. V.; McDermott, B.; Procopiou, P. A. *J. Chem. Soc. Perkin Trans. 1*, **2002**, 2237–2242.

14.   Bream, R. N.; Ley, S. V.; Procopiou, P. A. *Org. Lett.* **2002**, *4*, 3793–3796.

15.   Hett, R.; Stare, R.; Helquist, P. *Tetrahedron Lett.* **1994**, *35*, 9375–9378.

16.   Corey, E. J.; Bakshi, R. K.; Shibata, S. *J. Am. Chem. Soc.* **1987**, *109*, 5551–5553.

17.   Goswami, J.; Bezbaruah, R. L.; Goswami, A.; Borthakur, N. *Tetrahedron Asymmetry*, **2001**, *12*, 3343–3348.

18.   Procopiou, P. A.; Morton, G. E.; Todd, M.; Webb, G. *Tetrahedron Asymmetry*, **2001**, *12*, 2005–2008.

19.   Labelle, M.; Belley, M.; Gareau, Y.; Gauthier, J. Y.; Guay, D.; Gordon, R.; Grossman, S. G.; Jones, T. R.; Leblanc, Y.; McAuliffe, M.; McFarlane, C.; Masson, P.; Metters, K. M.; Ouimet, N.; Patrick, D. H.; Peichuta, H.; Rochette,

C.; Sawyer, N.; Xiang, Y. B.; Pickett, C. B.; Ford-Hutchinson, A. W.; Zamboni, R. J.; Young, R. N. *Bioorg. Med. Chem. Lett.* **1995**, *5*, 283–288.

20. Labelle, M.; Prasit, P.; Belley, M.; Blouin, M.; Champion, E.; Charette, L.; DeLuca, J. G.; Dufresne, R.; *et al. Bioorg. Med. Chem. Lett.* **1992**, *2*, 1141–1146.

21. Belly, M. L.; Leger, S.; Labelle, M.; Roy, P.; Xiang, Y. B.; Guay, D. US 5565473 (**1996**).

22. Young, R. N.; Zamboni, R.; Leger, S. US 4851409 (**1989**).

PROPERTY OF
SENECA COLLEGE
LIBRARIES
@ YORK CAMPUS